Simplified Building Design for Wind and Earthquake Forces

Simplified Building Design for Wind and Earthquake Forces

IIIIIIIIIIIIIIIIIIIIIIIIIIIIIIIIIIIIIIIIIIIIIIIIIIIIIIIIIIIIIIIIIIIIIIIIIIIIIIIIIIIIIIIIIII

James Ambrose

Professor of Architecture
University of Southern California

Dimitry Vergun

Consulting Structural Engineer
Lecturer in Architecture
University of Southern California

A Wiley-Interscience Publication

JOHN WILEY & SONS

New York • Chichester • Brisbane • Toronto

Library of Congress Cataloging in Publication Data:

Ambrose, James E
  Simplified building design for wind and earthquake
forces.

  "A Wiley-Interscience publication."
  Includes bibliographical references and index.
  1. Structural dynamics. 2. Wind-pressure.
3. Buildings—Aerodynamics. 4. Earthquake resistant
design. 5. Earthquakes and building. I. Vergun,
Dimitry, 1933-   joint author. II. Title.

TA654.A57   624.1'75   79-26660
ISBN 0-471-05013-X

Printed in the United States of America.

10 9 8 7 6 5 4 3 2 1

Preface

||||||||||||||||||||||||||||||||||||||||||||||||||||||||

This book is intended for those persons who, despite a lack of thorough training in structural engineering, wish to learn how to design ordinary building structures for wind and earthquake effects. Both of us have had considerable experience in dealing with such persons in our work as consulting structural engineers and as teachers of courses in structural design for architecture students. Although the book should also be valuable as an introduction to the subjects for those who intend to pursue a course of professional training in structural engineering, the scope of the material and the style of presentation has been developed with the former readers in mind.

The main portion of the book illustrates the analysis and design of the lateral bracing systems for 32 sample structures. These examples cover a range of size and type of structures that can be adequately designed with the simplified methods of equivalent static analysis. The principal reference used is the *Uniform Building Code,* portions of which are reprinted in the Appendix. The structural calculations for the examples can be understood by persons with a mathematical background limited to simple algebra and geometry and formal engineering study limited to statics, elementary strength of materials, and the analysis and design of simple structures of wood, steel, masonry, and reinforced concrete.

Although the engineering professions are moving steadily toward the exclusive use of international units (metrics), we have used only English units in this book simply because the references we have used all employ these units.

We are grateful to the International Conference of Building Officials, publishers of the *Uniform Building Code*, the Concrete Masonry Association of California and Nevada, Inryco, Inc., and the Simpson Company for permission to reprint materials from their publications. We are also grateful to our families for their patience and tolerance and their understanding of the time and effort required to develop this material. And finally, we are grateful to the editors and production personnel at John Wiley and Sons for their excellent work and invaluable assistance in helping to bring this material to successful publication.

JAMES AMBROSE
DIMITRY VERGUN

Los Angeles, California
March 1980

Contents

IIIIIIIIIIIIIIIIIIIIIIIIIIIIIIIIIIIIIIIIIIIIIIIIIIIIIIIIIIII

Simplified Building Design for Wind and Earthquake Forces

||||||||||||||||||||||||||||||||||||||||||||||||||

Introduction

IIIIIIIIIIIIIIIIIIIIIIIIIIIIIIIIIIIIIIIIIIIIIIIIIIIIIIIIIIIIIIII

The purpose of this book is to provide a source of study and reference for the topics of wind and earthquake effects as they pertain to the design of building structures. The treatment of these subjects is aimed at persons not trained in structural engineering but who have some background experience in the analysis and design of simple structures. Material presented includes the development of background topics, such as basic aspects of wind and earthquake effects and fundamentals of dynamic behavior, as well as the pragmatic considerations of design of structures for real situations.

As implied by the title, the scope of the work is limited. This limitation is manifested in the level of complexity of the problems dealt with and in the techniques used, principally with regard to the degree of difficulty in mathematical analysis and the sophistication of design methods. In order to set these limits we have assumed some specific minimal preparation by the reader, and individual readers should orient themselves with regard to these assumptions. For those with some lack of preparation, the list of references following Chapter 8 may be useful for supplementary study. For the reader with a higher capability in mathematics or a more intensive background in applied mechanics and structural analysis, this work may serve as a springboard to more rigorous study of the topics.

The majority of the mathematical work, especially that in the applied design examples in Chapters 6 and 7, is limited to relatively simple algebra and geometry. In the treatment of the fundamentals of dynamics and in the explanation of some of the formulas used in analysis and design it is occasionally necessary to use relationships from trigonometry, vector analysis, and calculus. The reader with this level of mathematical background will more fully appreciate the rational basis for the formulas, although their practical application will usually involve only simple algebra and arithmetic. Persons expecting to pursue the study of these topics beyond the scope of this book are advised to prepare themselves with work in mathematics that proceeds to the level of advanced calculus, partial differential equations, and matrix methods of analysis.

A minimal preparation in the topics of applied mechanics and structural analysis and design is assumed. This includes the topics of statics, elementary strength of materials, and the design of simple elements of wood, steel, and concrete structures for buildings. The general scope of the work in the design examples is limited to that developed in *Simplified Engineering for Architects and Builders* by Harry Parker (Ref. 2). When some of the examples involve the analysis of indeterminate structures, the work presented is done with simplified, approximate methods that should be reasonably well understood by the reader with the previously described minimal background. For a more rigorous and exact analysis of such problems, or for the study of more complex problems, the reader is advised to pursue a general study of the analysis of indeterminate structures.

A third area of assumed background knowledge is that of the ordinary materials and methods of building construction as practiced in the United States. It is assumed that the reader has a general familiarity with the ordinary processes of building construction and with the codes, standards, and sources of general data for structures of wood, steel, masonry, and concrete.

The major reference used for this work is the 1979 edition of the *Uniform Building Code* (Ref. 1), hereinafter called the *UBC*. The design examples in this book use the general requirements, the analytical procedures, and some of the specific data from this reference. Much of the material from the *UBC* that relates directly to problems of wind and earthquakes is reprinted in the Appendix to this book. It is recommended, however, that the reader have a copy of the entire code available because it contains considerable additional material pertinent to the use of specific materials, to structural design requirements in general, and to various problems of building planning and construction.

In real design situations individual buildings generally fall under the jurisdiction of a particular local code. Most large cities, many counties, and some states have their own individual codes. In many cases these codes are based primarily on one of the so-called "model" codes, such as the *UBC*, with some adjustments and additions for specific local conditions and practices. The reader who expects to work in a particular area is advised to obtain a copy of the code with jurisdiction in that area and to compare its provisions with those of the *UBC* as they are used in this work.

Building codes, including the *UBC*, are occasionally updated to keep them abreast of current developments in research, building practices, analytical and design techniques, and so on. The publishers of the *UBC* have generally followed a practice of issuing a new edition every three years. For reference in any real design work the reader is advised to be sure that the code he is using is the one with proper jurisdiction and is the edition currently in force. This precaution regarding use of dated materials applies also to other reference sources, such as handbooks, industry brochures, detailing manuals, and so on.

The best method for use of this book depends on the individual reader's background as well as his needs and intentions with regard to the topics of wind and earthquake design. The book is in general oriented about the presentation of the design examples in Chapters 6 and 7. Other parts of the book provide material that is considered to be background or supplementary to the work in these two chapters. The scope and purpose of these other portions of the book is as follows:

Chapter 1 contains a general discussion of dynamic effects on structures with emphasis on topics of concern in design for wind and earthquakes.

Chapter 2 discusses the general problems of lateral load resistance of building structures.

Chapter 3 discusses wind effects, including a summary of the *UBC* requirements.

Chapter 4 discusses earthquake effects, including a summary of the *UBC* requirements.

Chapter 5 discusses and illustrates the basic elements of typical lateral load-resisting systems for building structures, specific examples of which are shown in the various design cases in Chapters 6 and 7.

Chapter 8 contains discussions of a number of special problems.

A list of reference sources, including those used in this work and some that may serve to assist the reader in obtaining better background preparation or for further study, follows Chapter 8.

A glossary of words and terms generally encountered in discussions of structural design for wind and earthquakes is also provided. The Glossary, plus the Index, should be consulted whenever the reader encounters an unfamiliar word or term in this book.

Finally, the *Appendix* contains reprints of much material from the *UBC* and reprints or adaptations of materials from other sources that have been used in the examples in this book.

The reader with a general familiarity with the principles of dynamics, with the general nature and effects of wind and earthquakes, and with the fundamentals of analysis and design of ordinary structures may find it more useful to begin with a study of Chapter 5. From this introduction to the elements of simple lateral resistive systems he may then proceed to the examples in Chapters 6 and 7.

The reader will notice that, although design for earthquakes is usually more complicated than design for wind, Chapter 6 is the larger chapter. There is a practical reason for this. When the equivalent static load method is used, which is done exclusively in this book, the lateral load is generally dealt with in a similar manner, whether it is derived from wind effects or seismic effects. Thus, it is essentially the determination of the loading that differs in the two situations. Because the determination of wind loads is usually simpler, we have first developed relatively complete examples of analysis and design for wind, illustrating most of the simple, ordinary situations of structural materials and systems. Instead of following the same procedure in Chapter 7, we have concentrated on the determination of seismic loads, using many of the examples from Chapter 6 for reference to the design process once the loads have been found. This has been done essentially to shorten the work. We apologize to the readers whose interest is primarily in seismic design because this plan will make for somewhat clumsy reading, if Chapter 7 is read before studying Chapter 6. Therefore, we recommend to all readers that Chapters 5 and 6 be at least scanned before proceeding to Chapter 7.

Our use of the word *simplified* does not mean to imply that all design for wind and earthquakes can be reduced to simple methods. On the contrary, many problems in this area represent highly complex, and as yet far from fully understood, situations in structural design, situations that demand considerable seriousness, competency, and effort by professional engineers and researchers. We have deliberately limited the material in this book to that which we believe can be relatively easily understood and mastered by persons in the beginning stages of study of the design of structures. For those whose work will be limited to the relatively simple situations presented by the examples in this book, mastery of this material will provide useful working skills. For those who expect to continue their studies into more advanced levels of analysis and design, this material will provide a useful introduction.

1

Dynamic Effects on Building Structures

IIIIIIIIIIIIIIIIIIIIIIIIIIIIIIIIIIIIIIIIIIIIIIIIIIIIIII

Structural design for forces deriving from wind and earthquakes involves two special considerations. The first is that the principal application of the forces is in a horizontal direction, which makes them perpendicular (or lateral) to the direction of the force of gravity. Because most building structural systems are basically conceived in terms of their gravity resistance, designing for wind and earthquakes is often dealt with as the bracing of the gravity-resisting system against lateral forces. Although this practise is questionable as a basic design approach, it is nevertheless the way that most current design procedures have evolved.

Wind and earthquake forces are of course not limited to the horizontal direction. Because of aerodynamic effects and negative pressures (suction), wind forces may act up or down as well as in the direction of the horizontally moving air. Many roofs take leave of their supports during wind storms because they were not held down (against wind uplift) as well as held up (against gravity). Shock waves sent out by the ground faults that cause earthquakes result in motion in all directions, including up and down, of the ground surface and of buildings sitting on it. Because of the reserve strength resulting from the safety factors used for the

7

gravity design, the vertical effects are seldom critical, but they should not be totally ignored.

The second special consideration for wind and earthquake forces is that they are dynamic in nature. In order to understand these forces and the design for their resistance, it is necessary to have some appreciation for basic aspects of this dynamic character.

1.1 Static Versus Dynamic Effects

The word *static,* as used in mechanics, means that the forces are constant with respect to time. Eliminating the time variable permits an analysis that involves only the direct force properties of direction, sense, magnitude, and manner of application. Manner of application includes the location and the nature of dispersal of the forces—that is, whether they are concentrated at a point or distributed like a pressure. The basic unit of measurement for the magnitude of static forces, as well as for the response of structures to them, is a direct force unit in terms of gravity weight, such as ounces, pounds, or kilograms.

As shown in Figure 1.1, the effective strength of a structure in resisting static forces may be measured in these units. The structure's capacity for resistance may thus be simply compared to the applied force. A redundant (excess) capacity translates into a margin of safety; any shortcoming implies failure under something less than the full force of the load. For actual design it is usually necessary to take a more complex view, involving investigation for various types of internal force and the stresses and strains they produce. Nevertheless, the basic performance evaluation is the previous simple one: Does the structure take the load safely or not?

Analysis for deformation with static loads consists of the determination of the position of the loaded structure as compared to its unloaded position. Quantified deformations are generally measured in linear units. Evaluation of acceptability is done in terms of tolerable dimensional movement or shape change (curvature, twisting, bulging, etc.).

Time is not totally ignored in static analysis. Distinction is

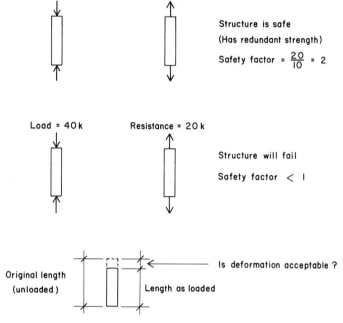

FIGURE 1.1 Evaluation of static force effects.

made between dead load (permanent) and live load (limited duration, random, etc.). There are also possibilities for time-dependent stress or strain behavior, such as fatigue in wood, creep in concrete, or progressive settlement in clay soils. However, these factors do not alter the basic nature of the static forces and their effects.

Analysis for dynamic loads and their effects on structures must include consideration of their time-dependent nature. Some of the mathematical implications of this are treated in the next section. The most significant point is that the performance of the structure must be dealt with in other than static terms for a truly dynamic analysis.

A simple example of these differences is shown in the illustrations in Figure 1.2. The graphs show typical stress and strain

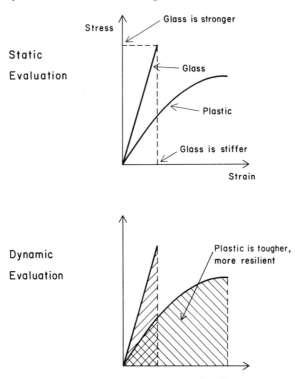

FIGURE 1.2 Dynamic versus static effects.

behavior for glass and acrylic plastic. The upper figure shows typical evaluations in static terms, the glass demonstrating both superior strength (higher stress resistance) and stiffness (less deformation). The lower figure shows an evaluation of an essential dynamic property, that of energy resistance. The energy capacity of the materials is represented by the area under each curve. This area represents the work done (product of force multiplied by distance, or stress multiplied by strain), and its total implies energy capacity because energy is the potential for doing work. Although not quite the whole story, this partly explains why a sheet of glass is more likely to fail under dynamic load than a sheet of plastic.

Besides energy capacity, the other most significant property for evaluation of dynamic resistance is the fundamental period of the structure in harmonic motion (vibration). Referring to the previous illustration, this means how long it takes the glass or the plastic to bounce back after a dynamic load of short duration. Comparison of the time for this bounce with the time of duration of the load will determine the relative dynamic effect of the loading.

One of the complexities of dealing with dynamic loads is that their effect on a structure is determined not only by their own dynamic properties but also by the nature of the dynamic behavior of the structure. Thus, the same loading will produce different effects on different structures. Some loads that are technically dynamic in nature tend to produce effects that are closer to being static in terms of a particular structure's response.

1.2 Basic Concepts of Dynamic Forces

A good lab course in physics should provide a reasonable understanding of the basic ideas and relationships involved in dynamic behavior. A better preparation is a course in engineering dynamics that focuses on the topics in an applied fashion, dealing directly with their applications in various engineering problems. The material in this section consists of a brief summary of basic concepts in dynamics that will be useful to those with a limited background and that will serve as a refresher for those who have studied the topics before.

The general field of dynamics may be divided into the areas of *kinetics* and *kinematics*. *Kinematics* deals exclusively with motion, that is, with time/displacement relationships and the geometry of movements. *Kinetics* adds the consideration of the forces that produce or resist motion.

Motion can be visualized in terms of a moving point, or in terms of the motion of a related set of points that constitute a body. The motion can be qualified geometrically and quantified dimensionally. In Figure 1.3 the point is seen to move along a path (its geometric character) a particular distance. The distance traveled by the point between any two separate locations on its path is

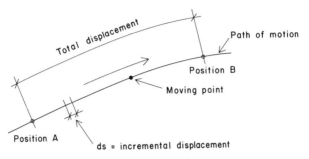

FIGURE 1.3   Kinematics of a moving point.

called *displacement*. The idea of motion is that this displacement occurs over time, and the general mathematical expression for the time/displacement function is

$$s = f(t)$$

Velocity is defined as the rate of change of the displacement with respect to time. As an instantaneous value, the velocity is expressed as the ratio of an increment of displacement (ds) divided by the increment of time (dt) elapsed during the displacement. Using the calculus, the velocity is thus defined as

$$v = \frac{ds}{dt}$$

That is, the velocity is the first derivative of the displacement.

If the displacement occurs at a constant rate with respect to time, it is said to have *constant velocity*. In this case the velocity may be expressed more simply without the calculus as

$$v = \frac{\text{total displacement}}{\text{total elapsed time}}$$

When the total velocity changes over time, its rate of change is called the *acceleration* (a). Thus, as an instantaneous change

$$a = \frac{dv}{dt} = \frac{d^2s}{dt^2}$$

That is, the acceleration is the first derivative of the velocity or the second derivative of the displacement with respect to time.

Except for the simplest cases, the derivation of the equations of motion for an object generally require the use of the calculus in the operation of these basic relationships. Once derived, however, motion equations are generally in algebraic form and can be used without the calculus for application to problems. An example is the set of equations that describe the motion of a free-falling object acted on by the earth's gravity field. Under idealized conditions (ignoring air friction, etc.) the distance of fall from a rest position will be

$$s = f(t) = 16.1\, t^2 \qquad (s \text{ in ft}, t \text{ in sec})$$

This equation indicates that the rate of fall (the velocity) is not a constant but increases with the elapsed time, so that the volocity at any instant of time may be expressed as

$$v = \frac{ds}{dt} = \frac{d(16.1\, t^2)}{dt} = 32.2\, t \qquad (v \text{ in ft/sec})$$

and the acceleration as

$$a = \frac{dv}{dt} = \frac{d(32.2\, t)}{dt} = 32.2 \text{ ft/sec}^2$$

which is the acceleration of gravity.

Kinematics also includes the study of the various forms of motion: translation, rotation, plane motion, motion of deformable bodies, and so on. A study of the mechanics of motion is very useful in the visualization of the deformation of a structure by static as well as dynamic forces.

As stated previously, kinetics includes the additional consideration of the forces that cause motion. This means that in addition to the variables of displacement and time, we must consider the mass of the moving objects. From Newtonian physics the simple definition of mechanical force is

$$F = ma = \text{mass} \times \text{acceleration}$$

Mass is the measure of the property of inertia, which is what causes an object to resist change in its state of motion. The more common term for dealing with mass is *weight*, which is a force defined as

$$W = mg$$

where g is the constant acceleration of gravity (32.2 ft/sec^2).

Weight is literally a dynamic force, although it is the standard means of measurement of force in statics, when the velocity is assumed to be zero. Thus, in static analysis we express forces simply as

$$F = W$$

and in dynamic analysis, when using weight as the measure of mass, we express force as

$$F = ma = \frac{W}{g}a$$

If a force moves an object, work is done. *Work* is defined as the product of the force multiplied by the displacement (distance traveled). If the force is constant during the displacement, work may be simply expressed as

$$w = Fs = \text{force} \times \text{total distance traveled}$$

If the force varies with time, the relationship is more generally expressed with the calculus as

$$w = \int_{s_2}^{s_1} F_t ds$$

indicating that the displacement is from position s_1 to positions s_2, and the force varies in some manner with respect to time.

Figure 1.4 illustrates these basic relationships. In dynamic analysis of structures the dynamic "load" is often translated into work units in which the distance traveled is actually the deformation of the structure.

Energy may be defined as the capacity to do work. Energy exists in various forms: heat, mechanical, chemical, and so on. For structural analysis the concern is with mechanical energy,

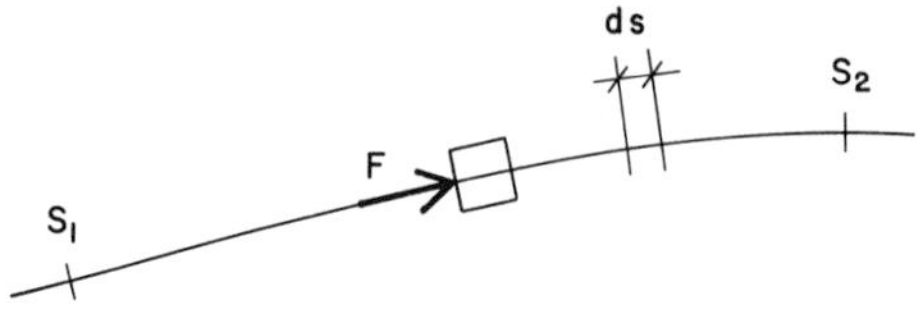

$$\text{Work} = \int_{S_1}^{S_2} F_t \, ds \quad (F \text{ variable with time})$$

$$= F(S_2 - S_1) \quad (F \text{ constant with time})$$

FIGURE 1.4   Kinetics of a moving object.

which occurs in one of two forms. *Potential energy* is stored energy, such as that in a compressed spring or an elevated weight. Work is done when the spring is released or the weight is dropped. *Kinetic energy* is possessed by bodies in motion, work is required to change their state of motion, that is, to slow them down or speed them up. (See Figure 1.5.)

In structural analysis energy is considered to be indestructible, that is, it cannot be destroyed, although it can be transferred or transformed. The potential energy in the compressed spring can be transferred into kinetic energy if the spring is used to propel an object. In a steam engine the chemical energy in the fuel is transformed into heat and then into pressure of the steam and finally into mechanical energy delivered as the engine's output.

An essential idea is that of the conservation of energy, which is a statement of its indestructibility in terms of input and output. This idea can be stated in terms of work by saying that the work done on an object is totally used and that it should therefore be equal to the work accomplished plus any losses due to heat, air friction, and so on. In structural analysis we make use of this concept by using a "work equilibrium" relationship similar to the static force equilibrium relationship. Just as all the forces must be in balance for static equilibrium, so the work input must equal the work output (plus losses) for "work equilibrium."

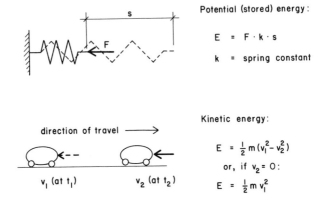

Potential (stored) energy:

$$E = F \cdot k \cdot s$$

$$k = \text{spring constant}$$

direction of travel ⟶

Kinetic energy:

$$E = \tfrac{1}{2} m (v_1^2 - v_2^2)$$

or, if $v_2 = 0$:

$$E = \tfrac{1}{2} m v_1^2$$

v_1 (at t_1) v_2 (at t_2)

FIGURE 1.5   Forms of mechanical energy.

A special kinematic problem of major concern in structural analysis for dynamic effects is that of *harmonic motion*. The two elements generally used to illustrate this type of motion are the swinging pendulum and the bouncing spring. Both the pendulum and the spring have a neutral position where they will remain at rest in static equilibrium. If one displaces either of them from this neutral position, by pulling the pendulum sideways or compressing or stretching the spring, they will tend to move back to the neutral position. Instead of stopping at the neutral position, however, they will be carried past it by their momentum to a position of displacement in the opposite direction. This sets up a cyclic form of motion (swinging of the pendulum; bouncing of the spring) that has some basic characteristics.

Figure 1.6 illustrates the typical motion of a bouncing spring. Using the calculus and the basic motion and force equations, the displacement/time relationship may be derived as

$$s = A \cos Bt$$

The cosine function produces the basic form of the graph, as shown in Figure 1.6. The maximum displacement from the neutral position is called the *amplitude*. The time elapsed for one full cycle is called the *period*. The number of full cycles in a given unit of time is called the *frequency* (usually expressed in cycles per

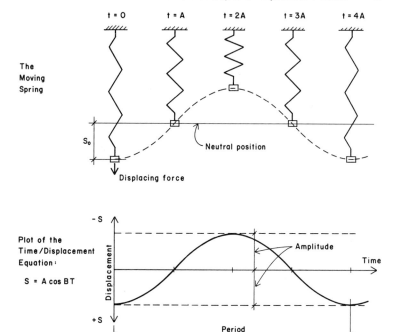

FIGURE 1.6   Harmonic motion of a bouncing spring.

second) and is equal to the inverse of the period. Every object subject to harmonic motion has a fundamental period (also called natural period), which is determined by its weight, stiffness, size, and so on.

Any influence that tends to reduce the amplitude in successive cycles is called a *damping effect*. Heat loss in friction, air resistance, and so on are natural damping effects. Shock absorbers, counterbalances, cushioning materials, and other devices can also be used to damp the amplitude. Figure 1.7 shows the form of a damped harmonic motion, which is the normal form of most such motions, because perpetual motion is not possible without a continuous reapplication of the original displacing force.

Resonance is the effect produced when the displacing effort is itself harmonic with a cyclic nature that corresponds with the

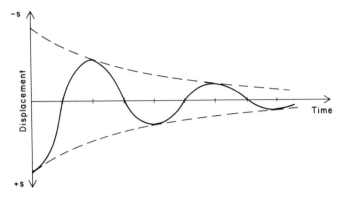

FIGURE 1.7   Form of a damped harmonic motion.

period of the impelled object. An example is someone bouncing on a diving board in rhythm with the board's fundamental period, thus causing a reinforcement, or amplification, of the board's free motion. This form of motion is illustrated in Figure 1.8. Unrestrained resonant effects can result in intolerable amplitudes, producing destruction or damage of the moving object or its supports. A balance of damping and resonant effects can sometimes produce a constant motion with a flat profile of the amplitude peaks.

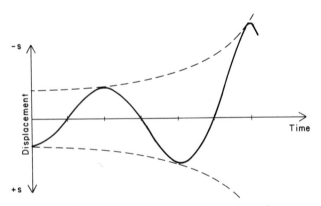

FIGURE 1.8   Form of a resonant harmonic motion.

Loaded structures tend to act like springs. Within the elastic stress range of the materials, they can be displaced from a neutral (unloaded) position and, when released, will go into a form of harmonic motion. The fundamental period of the structure as a whole, as well as the periods of its parts, are major properties that affect responses to dynamic loads.

1.3 Dynamic Effects on Structures

Load sources that involve motion, such as wind, earthquakes, walking people, moving vehicles, and vibrating heavy machinery, have the potential to cause dynamic effects on structures. Analyzing for their effects requires consideration of essential dynamic properties of the structure. These properties are determined by the size, weight, relative stiffness, fundamental period, type of support, and degree of elasticity of the materials of the structure and by various damping influences that may be present.

Dynamic load sources deliver an energy load to the structure that may be in the form of an impact, such as that caused by the moving air bumping into the stationary building. In this case the energy load is derived from the kinetic energy of the moving air, which is a product of its mass and velocity. In the case of an earthquake, or the vibration of heavy machinery, the load source is not a force as such but, rather, something that induces motion of the structure, in which case the mass of the building is actually the load source.

An important point to note is that the effects of a dynamic load on a structure are determined by the structure's response as well as by the nature of the load. Thus the same dynamic load can produce different effects in different structures. Two buildings standing side by side can have significantly different responses to the same earthquake shock if they have major differences in their dynamic properties.

Dynamic effects on structures may be of several types. Some of the principal effects are the following:

Total energy load is the balance between the peak magnitude of the load and the maximum work required by the structure and is known as the *work equilibrium concept.* Work done to the structure by the load equals the work done by the structure in resisting the load.

Unstabilizing effects occur if the dynamic load produces a stability failure of the structure. Thus, a free-standing wall may topple over, an unbraced post and beam system may collapse sideways, and so forth, because of the combined effects of gravity and the dynamic load.

Harmonic effects of various types may be set up in the structure, especially if the load source is cyclic in nature, such as the footsteps of marching troops. Earthquake motions are basically cyclic, in the form of vibration or shaking of the surface of the ground. Relations between these motions and the harmonic properties of the structure can result in various effects, such as the flutter of objects at a particular wind velocity, the resonant bouncing of floors, and the resonant reinforcing of the swaying of buildings during an earthquake.

Failure under repeated loadings can result in some cases when structures literally use up their dynamic resistance. The structures may successfully resist a single peak load of some dynamic effort, only to fail later under a similar, or even smaller, loading. This failure is usually due to the fact that the first loading used up some degree of structural failure, such as ductile yielding or brittle cracking, which absorbed enough energy to prevent total failure but was a one-time usable strength only.

A major consideration in design for dynamic loads is what the response of the structure means to the building as a whole. Thus, although the structure may remain intact, that may be only a minor accomplishment if there is significant damage to the building as a whole. A high-rise building may swing and sway in an earthquake without there being any significant damage to the structure, but if the occupants are tossed about, the ceilings fall, the windows shatter, the partitions and curtain walls collapse, the plumbing bursts, and the elevators derail, it can hardly be said that the building was adequately designed.

In many cases analysis and design for dynamic effects are not done by working directly with the dynamic relationships but simply by using recommendations and rules of thumb that have been established by experience. Some testing or theoretical analysis may have helped in deriving ideas or data, but much of what is used is based on the observations and records from previous disasters. Even when actual calculations are performed, they are mostly done with data and relationships that have been translated into simpler static terms—so-called equivalent static analysis and design. The reasons for this practise have to do frankly with the degree of complexity of dynamic analysis. Even with the use of programmable calculators or computers, the work is quite laborious in all but the simplest of situations.

1.4   Design for Equivalent Static Effects

Use of equivalent static effects essentially permits simpler analysis and design by eliminating the complex procedures of dynamic analysis. To make this possible the load effects and the structure's responses must be translated into static terms.

For wind load the primary translation consists of converting the kinetic energy of the wind into an equivalent static pressure, which is then treated in a manner similar to that for a distributed gravity load. Additional considerations are made for various aerodynamic effects, such as ground surface drag, building shape, and suction, but these do not change the basic static nature of the work.

For earthquake effects the primary translation consists of establishing a hypothetical horizontal static force that is applied to the structure to simulate the effects of sideward motions during ground movements. This force is calculated as some percentage of the dead weight of the building, which is the actual source of the kinetic energy loading once the building is in motion—just as the weight of the pendulum and the spring keep them moving after the initial displacement and release. The specific percentage used is determined by a number of factors, including some of the dynamic response characteristics of the structure.

An apparently lower safety factor is used when designing for the effects of wind and earthquake because an increase of one-third is permitted in allowable stresses. This is actually not a matter of a less-safe design but is merely a way of compensating for the fact that one is actually adding static (gravity) effects and *equivalent* static effects. The total stresses thus calculated are really quite hypothetical because in reality one is adding static strength effects to dynamic strength effects, in which case 2 + 2 does not necessarily make 4.

Regardless of the number of modifying factors and translations, there are some limits to the ability of an equivalent static analysis to account for dynamic behavior. Many effects of damping and resonance cannot be accounted for. The true energy capacity of the structure cannot be accurately measured in terms of the magnitudes of stresses and strains. There are some situations, therefore, in which a true dynamic analysis is desirable, whether it is performed by mathematics or by physical testing. These situations are actually quite rare, however. The vast majority of building designs present situations for which a great deal of experience exists. This experience permits generalizations on most occasions that the potential dynamic effects are really insignificant or that they will be adequately accounted for by design for gravity alone or with use of the equivalent static techniques.

2

Lateral Load Resistance
of Buildings

IIIIIIIIIIIIIIIIIIIIIIIIIIIIIIIIIIIIIIIIIIIIIIIIIIIIIIIIIIIII

To understand how a building resists the lateral load effects of
wind and seismic force it is necessary to consider the manner of
application of the forces and then to visualize how these forces
are transferred through the lateral resistive structural system and
into the ground. Although it is well to bear in mind that the forces
are basically dynamic in nature, we will deal with them as equivalent static forces in this discussion.

2.1 Application of Lateral Forces

The application of wind forces to a closed building is in the form
of pressures applied normal to the exterior surfaces of the building. As has been mentioned previously, these forces may be
either inward (called *positive pressure*) or outward (called *negative* or *suction pressure*). The shape of the building and the
direction of the wind primarily determine the nature of distribution of pressures on the various exterior surfaces of the building.
The total effect on the building is usually determined by considering the vertical profile, or silhouette, of the building as a single
vertical plane surface at right angles to the wind direction. A

direct pressure is assumed to act on this plane, simulating the combined effects of push, drag, and suction on the actual surfaces of the building.

Figure 2.1 shows a simple box-shaped building under the effect of wind normal to one of its flat sides. The lateral resistive structure that responds to this loading consists of the following:

Wall surface elements on the windward side are assumed to take the total wind pressure on the building and are calculated as previously described. In this structure they span vertically, transferring the load to the horizontal structure.

Roof and floor diaphragms, considered as rigid planes, receive an edge loading from the exterior wall and distribute it to the vertical bracing elements.

Vertical diaphragms or shear walls, when assumed to be the two end walls parallel to the direction of the wind, receive the load as a linearly distributed load from the edges of the horizontal diaphragms. Acting as vertical cantilevers, these walls transfer the loads to their supports, in this case the wall foundations.

The foundation is where the buck stops. The foundation must resist the overturning (toppling) and horizontal shear at the base of the shear walls. This resistance can be developed only by the dead weight of the foundation, the vertical soil pressure and sliding resistance on the bottom of the footings, and the lateral earth pressure on the side of the foundation on the leeward side (opposite from the wind).

The basic functions of these elements of the lateral resistive system are shown in Figure 2.2. The exterior wall functions as a simple spanning element, loaded by a uniformly distributed pressure normal to its surface and delivering a reaction force to its supports. In most cases, even though the wall may be continuous through several stories, it is considered as a simple span at each story level, thus delivering half of its load to each support. Referring to Figure 2.1, this means that the upper wall delivers half of its load to the roof edge and half to the edge of the second floor.

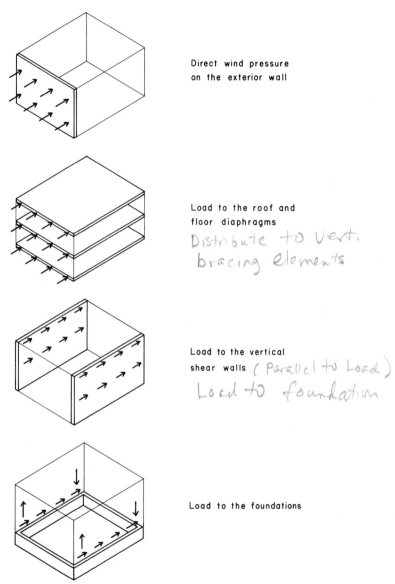

Direct wind pressure
on the exterior wall

Load to the roof and
floor diaphragms

Distribute to Vert. bracing elements

Load to the vertical
shear walls *(Parallel to Load)*

Load to foundation

Load to the foundations

FIGURE 2.1 Propagation of wind force in a box building.

25

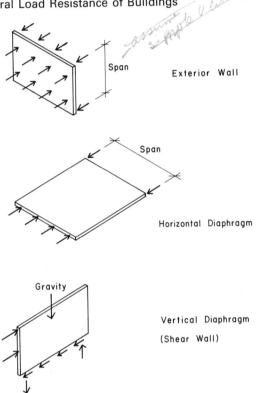

assume 2. apple V est span

Exterior Wall

Horizontal Diaphragm

Vertical Diaphragm
(Shear Wall)

FIGURE 2.2   Basic functions of elements of a box system.

The lower wall delivers half of its load to the second floor and half to the first floor.

This may be a somewhat simplistic view of the function of the walls themselves, depending on their construction. If they are framed walls with windows or doors, there may be many internal load transfers within the wall. Usually, however, the external load delivery to the horizontal structure will be as described.

The roof and second floor diaphragms function as spanning elements loaded by the edge forces from the exterior wall and spanning between the end shear walls, thus producing a bending that develops tension on the leeward edge and compression on the windward edge. It also produces shear in the plane of the diaphragm that becomes a maximum at the end shear walls. In most cases the shear is assumed to be taken by the diaphragm, but the tension and compression forces due to bending are transferred to framing at the diaphragm edges. The means of achieving this transfer depends on the materials and details of the construction.

The end shear walls act as vertical cantilevers that also develop shear and bending. The total shear in the upper story is equal to the edge load from the roof. The total shear in the lower story is the combination of the edge loads from the roof and second floor. The total shear force in the wall is delivered at its base in the form of a sliding friction between the wall and its support. The bending caused by the lateral load produces an overturning effect at the base of the wall as well as the tension and compression forces at the edges of the wall. The overturning effect is resisted by the stabilizing effect of the dead load on the wall. If this stabilizing moment is not sufficient, a tension tie must be made between the wall and its support.

If the first floor is attached directly to the foundations, it may not actually function as a spanning diaphragm but rather will push its edge load directly to the leeward foundation wall. In any event, it may be seen in this example that only three-quarters of the total wind load on the building is delivered through the upper diaphragms to the end shear walls.

This simple example illustrates the basic nature of the propagation of wind forces through the building structure, but there are many other possible variations with more complex building forms or with other types of lateral resistive structural systems. Some of these variations are discussed in the next section of this chapter and in Chapter 5.

Seismic loads are actually generated by the dead weight of the building construction. In visualizing the application of seismic forces, we look at each part of the building and consider its weight as a horizontal force. The weight of the horizontal structure, although actually distributed throughout its plane, may usually be dealt with in a manner similar to the edge loading caused by wind. In the direction normal to their planes, vertical walls will be loaded and will function structurally in a manner similar to that for direct wind pressure. The load propagation for the box-shaped building in Figure 2.1 will be quite similar for both wind and seismic forces.

If a wall is reasonably rigid in its own plane, it tends to act as a vertical cantilever for the seismic load in the direction parallel to its surface. Thus, in the example building, the seismic load for the roof diaphragm would usually be considered to be caused by the weight of the roof and ceiling construction plus only those walls whose planes are normal to the direction being considered. These different functions of the walls are illustrated in Figure 2.3. If this assumption is made, it will be necessary to calculate a separate seismic load in each direction for the building.

For determination of the seismic load, it is necessary to consider all elements that are permanently attached to the structure. Ductwork, lighting and plumbing fixtures, supported equipment, signs, and so on will add to the total dead weight for the seismic load. In buildings such as storage warehouses and parking garages it is also advisable to add some load for the building contents.

2.2 Types of Lateral Resistive Systems

The building in the previous example illustrates one type of lateral resistive system: the box, or panelized, system. As shown in Figure 2.4, the general types of systems are those discussed in the following paragraphs.

The Box, or Panelized, System.   The box, or panelized, system is usually of the type shown in the previous example, consisting of some combination of horizontal and vertical planar elements. Actually, most buildings use horizontal diaphragms, simply because the existence of roof and floor construction provides them as a matter of course. The other types of systems usually consist of variations of the vertical bracing elements. An occasional exception is a roof structure that must be braced by trussing or other means when there are a large number of roof openings or a roof deck with little or no diaphragm strength.

Roof & Floor = Horiz Diaphragms

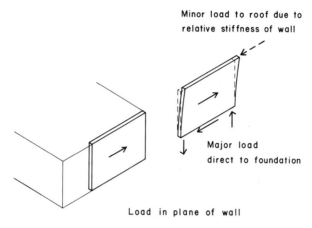

Load in plane of wall

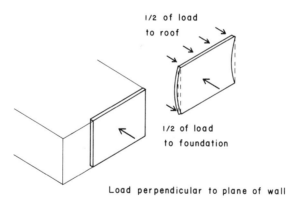

Load perpendicular to plane of wall

FIGURE 2.3 Seismic loads caused by wall weight.

Internally Braced Frames. The typical assemblage of post and beam elements is not inherently stable under lateral loading unless the frame is braced in some manner. Shear wall panels may be used to achieve this bracing, in which case the system functions as a box, even though there is a frame structure. It is also possible, however, to use diagonal members, X-bracing, knee braces, struts, and so on to achieve the necessary stability of

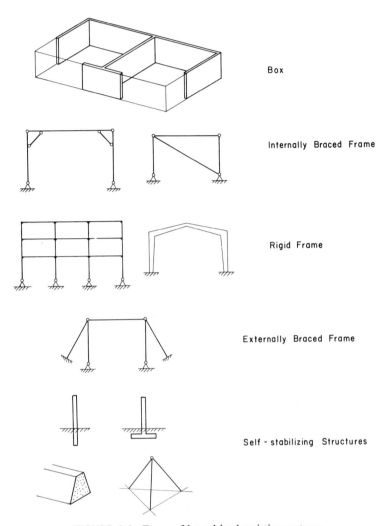

Box

Internally Braced Frame

Rigid Frame

Externally Braced Frame

Self - stabilizing Structures

FIGURE 2.4 Types of lateral load resistive systems.

30

the rectangular frame. The term *braced frame* usually refers to these techniques.

Rigid Frames. Although the term *rigid frame* is a misnomer, since this technique usually produces the most *flexible* lateral resistive system, the term refers to the use of moment resistive joints between the elements of the frame.

Externally Braced Frames. The use of guys, struts, buttresses, and so on that are applied externally to the structure or the building results in externally braced frames.

Self-stabilizing Elements and Systems. Retaining walls, flagpoles, pyramids, tripods, and so on, in which stability is achieved by the basic form of the structure are examples of self-stabilizing elements and systems.

Each of these systems has variations in terms of materials, form of the parts, details of construction, and so on. These variations may result in different behavior characteristics, although each of the basic types has some particular properties. An important property is the relative stiffness or resistance to deformation, which is of particular concern in evaluating energy effects, especially for response to seismic loads. A box system with diaphragms of poured-in-place concrete is usually very rigid, having little deformation and a short fundamental period. A multistory rigid frame of steel, on the other hand, is usually quite flexible and will experience considerable deformation and have a relatively long fundamental period. In seismic analysis these properties are used to modify the percentage of the dead weight that is used as the equivalent static load to simulate the seismic effect.

Elements of the building construction developed for the gravity load design, or for the general architectural design, may become natural elements of the lateral resistive system. Walls of the proper size and in appropriate locations may be theoretically functional as shear walls. Whether they can actually serve as such will depend on their construction details, on the materials used, on their length-to-height ratio, and on the manner in which they are attached to the other elements of the system for load transfer.

It is also possible, of course, that the building construction developed only for gravity load resistance and architectural planning considerations may *not* have the necessary attributes for lateral load resistance, thus requiring some replanning or the addition of structural elements.

Many buildings consist of mixtures of the basic types of lateral resistive systems. Walls existing with a frame structure, although possibly not used for gravity loads, can still be used to brace the frame for lateral loads. Shear walls may be used to brace a building in one direction whereas a braced frame or rigid frame is used in the perpendicular direction. Multistory buildings occasionally have one type of system, such as a rigid frame, for the upper stories and a different system, such as a box system or braced frame, for the lower stories to reduce deformation and take the greater loads in the lower portion of the structure.

In many cases it is neither necessary nor desirable to use every wall as a shear wall or to brace every bay of the building frame. The illustrations in Figure 2.5 show various situations in which the lateral bracing of the building is achieved by partial bracing of the system. This procedure does require that there be some load-distributing elements, such as the roof and floor diaphragms, horizontal struts, and so on, that serve to tie the unstabilized portions of the building to the lateral resistive elements.

There is a possibility that some of the elements of the building construction that are not intended to function as bracing elements may actually end up taking some of the lateral load. In frame construction, surfacing materials of plaster, drywall, wood paneling, masonry veneer, and so on may take some lateral load even though the frame is braced by other means. This is essentially a matter of relative stiffness, although connection for load transfer is also a consideration. What can happen in these cases is that the stiffer finish materials take the load first, and if they are not strong enough, they fail and the intended bracing system then goes to work. Although collapse may not occur, there can be considerable damage to the building construction as a result of the failure of the supposed nonstructural elements.

The choice of the type of lateral resistive system must be related to the loading conditions and to the behavior characteris-

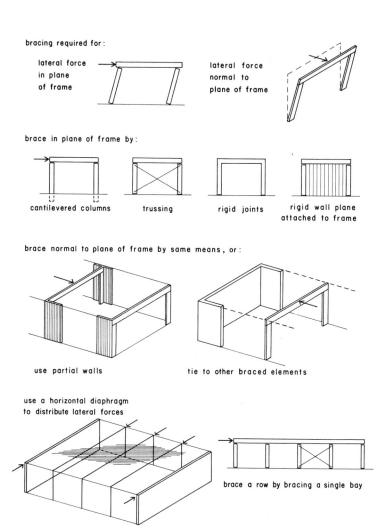

FIGURE 2.5 Bracing of framed structures for lateral loads.

tics required. It must also, however, be coordinated with the design for gravity loads and with the architectural planning considerations. Many design situations allow for alternatives, although the choice may be limited by the size of the building, by code restrictions, by the magnitude of lateral loads, by the desire for limited deformation, and so on.

2.3 Design Considerations

The design of the lateral load resistive system for a building involves a great number of factors. The principal considerations are the following.

Determination of the Loading. Determining the loading is usually established by the satisfaction of the requirements of the building code with jurisdiction. Critical load values, as well as various requirements for the form of the structural analysis and design, are determined by the degree of local concern for extremes of wind storms or earthquakes. This concern is primarily based on the history of disasters in the area.

Selection and Planning of the Lateral Resistive System for the Building. As previously discussed, this selection and planning must be coordinated with the gravity load design and the architectural design in general. In some cases the design for lateral loads may be a major factor in establishing the building form and detail, in selection of materials, and so on. In other cases it may consist essentially of assuring the proper construction of ordinary elements of the construction.

Detailed Analysis and Design of the Elements of the Lateral Resistive System. With the loading established and the system defined, the performance of the individual parts and of the system as a whole must be investigated. An important aspect of the investigation is the complete following through of the loads from their origin to their final resolution in the ground. With the internal forces and stresses determined the design of the parts of the system is usually a matter of routine, using code specifications and data from the code or from other reference sources.

Development of Structural Construction Details and Specifications. Such development essentially constitutes the documentation of the design and is a task of major importance. A thorough analytical investigation and complete set of structural calculations will be useless unless the results are translated into directives usable by the builders of the building.

Convincing the Authorities Who Grant Building Permits that the Structure is Adequate. In most cases someone employed or retained as a consultant by the code enforcing body will review the structural calculations and the construction drawings and specifications for compliance with the local code requirements and with acceptable practices of design and construction. Although it should be expected that a competent and thorough design effort will receive a good review, there is usually some room for individual judgment and personal preference, so that the potential exists for some conflict between the designer and the reviewer.

In some ways the easiest part of this process is that of the structural analysis and design. The analysis and design may be laborious if the building is large or complex, but it is usually routine in nature, with considerable information and guidance available from codes, texts, industry brochures, and so on. Some degree of training in engineering mechanics and basic structural analysis and design is necessary, but most of the work is ''cookbook'' in nature.

Determination of the lateral loads is also reasonably simple, at least with the use of the equivalent static load methods. One possible complication is due to the fact that the structure must be defined in some detail before the loads can be determined and their propagation through the system investigated, which is somewhat like needing to know the answer before you can formulate the question. Some things must be known about the structure before you can analyze its behavior. As a result, the early stages of design often consist of some guessing and trying—approximating a structure and then analyzing it to see if it works. This process is easier, of course, when the designer has worked on similar problems before, or if he has the results of previous similar designs as a basis for a more educated first guess. The more difficult aspects of design for lateral loads are the development of the basic systems for the lateral resistive structure and the development of the necessary construction drawings and specifications to assure proper construction. This work requires considerable understanding of the problems of the building design and construction in general because decisions about the basic structural scheme and some of the details of the structure may have considerable influence on the general form and detail of the building and on the economics and general feasibility of the construction.

Getting a building permit is somewhat like going to court. The permit-granting authorities represent the prosecution, judge, and jury. The structural designer and architect represent the counsel for the plaintiff. The burden of proof is on the plaintiff. As with any court case, it helps to have a good case to make (no code requirements violated; no questionable design or construction procedures) and to present it well (clear, complete structural calculations and construction drawings and specifications).

3

Wind Effects on Buildings

||||||||||||||||||||||||||||||||||||||||||||||||||||||

Wind is moving air. The air has a particular mass (density or weight) and moves in a particular direction at a particular velocity. It thus has kinetic energy of the form expressed as

$$E = \frac{1}{2} mv^2$$

When the moving fluid air encounters a stationary object, there are several effects that combine to exert a force on the object. The nature of this force, the many variables that affect it, and the translation of the effects into criteria for structural design are dealt with in this chapter.

3.1 Wind Conditions

The wind condition of concern for building design is primarily that of a wind storm, specifically high-velocity, ground-level winds. These winds are generally associated with one of the following situations.

Tornadoes. Tornadoes occur with some frequency in the Midwest and occasionally in other parts of the United States. In coastal areas they are usually the result of ocean storms that wander ashore. Although the most violent effects are at the center of the storm, high-velocity winds in a large surrounding area often accompany these storms. In any given location the violent winds are usually short in duration as the tornado dissipates or passes through the area.

Hurricanes.   Whereas tornadoes tend to be relatively short-lived (a few hours at most), hurricanes can sustain storm wind conditions for several days. Hurricanes occur with some frequency in the Atlantic and Gulf coastal areas of the United States. Although they originate and develop their greatest fury over the water, they often stray ashore and can move some distance inland before dissipating. As with tornadoes, the winds of highest velocity occur at the eye of the hurricane, but major winds can develop in large surrounding areas, often affecting coastal areas some distance inland even when the hurricane stays at sea.

Local Peculiar Wind Conditions.   An example of wind conditions peculiar to one locality are the Santa Ana winds of Southern California. These winds are recurrent conditions caused by the peculiar geographic and climatological conditions of an area. They can sometimes result in local wind velocities of the level of those at the periphery of tornadoes and hurricanes and can be sustained for long periods.

Sustained Local Wind Conditions.   Winds that occur at great elevations above sea level are an example of sustained local wind conditions. Such winds may possibly never reach the extremes of velocity of storm conditions, but they can require special consideration because of their enduring nature.

Local and regional meteorological histories are used to predict the degree of concern for or likelihood of critical wind conditions in a particular location. Building codes establish minimum design requirements for wind based on this experience and the statistical likelihood it implies. The map in the *UBC*, Figure 4 (see the Appendix), shows the variation of critical wind conditions in the United States.

Of primary concern in wind evaluation is the maximum velocity that is achieved by the wind. Maximum velocity usually refers to a sustained velocity and not to gust effects. A gust is essentially a pocket of higher velocity wind within the general moving fluid air mass. The resulting effect of a gust is that of a brief increase, or surge, in the wind velocity, usually of not more than 15% of the sustained velocity and for only a fraction of a second in duration. Because of both its higher velocity and its slamming effect, the gust actually represents the most critical effect of the wind in most cases.

Winds are measured regularly at a large number of weather stations. The standard measurement is taken at an elevation of 30 ft above the surrounding terrain, which provides a fixed reference with regard to the drag effects of the ground surface. The graph in Figure 3.1 shows the correlation between wind velocity and various wind conditions. The curve on the graph is a plot of the general equation used to relate wind velocity to equivalent static pressure on buildings, as discussed in the next section.

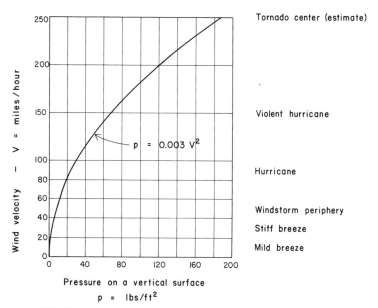

FIGURE 3.1 Relation of wind velocity to pressure on buildings.

Although wind conditions are usually generalized for a given geographic area, they can vary considerably for specific sites because of the nature of the surrounding terrain, of landscaping, or of nearby structures. Each individual building design should consider the possibilities of these localized site conditions.

3.2 General Wind Effects

The effects of wind on stationary objects in its path can be generalized as in the following discussions (see Figure 3.2).

Direct Positive Pressure.   Surfaces facing the wind and perpendicular to its path receive a direct impact effect from the moving mass of air, which generally produces the major portion of force on the object, unless it is highly streamlined in form.

Aerodynamic Drag.   Because the wind does not stop upon striking the object but flows around it like a fluid, there is a drag effect on surfaces that are parallel to the direction of the wind. These surfaces may also have inward or outward pressures exerted on them, but it is the drag effect that adds to the general force on the object in the direction of the wind path.

Negative Pressure.   On the leeward side of the object (opposite from the wind direction) there is usually a suction effect, consisting of pressure outward on the surface of the object. By comparison to the direction of pressure on the windward side, this is called *negative pressure.*

These three effects combine to produce a net force on the object in the direction of the wind that tends to move the object along with the wind. In addition to these there are other possible effects on the object that can occur due to the turbulence of the air or to the nature of the object. Some of them are as follows.

Rocking Effects.   During wind storms, the wind velocity and its direction are seldom constant. Gusts and swirling winds are ordinary, so that an object in the wind path tends to be buffeted, rocked, flapped, and so on. Objects with loose parts, or with connections having some slack, or with highly flexible surfaces

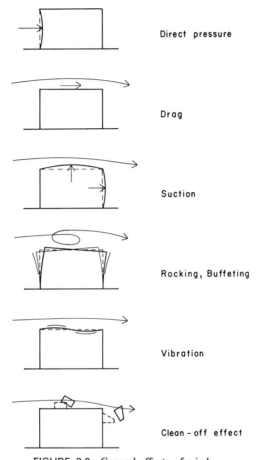

Direct pressure

Drag

Suction

Rocking, Buffeting

Vibration

Clean - off effect

FIGURE 3.2 General effects of wind.

(such as fabric surfaces that are not taut) are most susceptible to these effects.

Harmonic Effects. Anyone who plays a wind instrument appreciates that wind can produce vibration, whistling, flutter, and so on. These effects can occur at low velocities as well as with wind storm conditions. This is a matter of some match between the velocity of the wind and the natural period of vibration of the object or of its parts.

Clean-Off Effect. The friction effect of the flowing air mass tends to smooth off the objects in its path. This fact is of particular concern to objects that protrude from the general mass of the building, such as canopies, parapets, chimneys, and signs.

The critical condition of individual parts or surfaces of an object may be caused by any one, or some combination, of these effects. Damage can occur locally or be total with regard to the object. If the object is resting on the ground, it may be collapsed or may be slid, rolled over, or lifted from its position. Various aspects of the wind, of the object in the path of the wind, or of the surrounding environment determine the critical wind effects. With regard to the wind itself some considerations are the following:

The magnitude of sustained velocities.

The duration of high-level velocities.

The presence of gust effects, swirling, and so on.

The prevailing direction of the wind (if any).

With regard to objects in the path of the wind some considerations are the following:

The size of the object (relates to the relative effect of gusts, to variations of pressure above ground level, and so on).

The aerodynamic shape of the object (determines the critical nature of drag, suction, uplift, and so on).

The fundamental period of vibration of the object or of its parts.

The relative stiffness of surfaces, tightness of connections, and so on.

With regard to the environment, possible effects may result from the sheltering or funneling caused by ground forms, landscaping, or adjacent structures. These effects may result in an increase or reduction of the general wind effects or in turbulence to produce a very unsteady wind condition.

The actual behavior of an object during wind storm conditions can be found only by subjecting it to a real wind situation. Wind tunnel tests in the laboratory are also useful, and because we can create the tests more practically on demand, they have provided much of the background for data and procedures used in design.

3.3 Critical Wind Effects on Buildings

The major effects of wind on buildings can be generalized to some degree because we know a bracketed range of characteristics that cover the most common conditions. Some of the general assumptions made are as follows:

Most buildings are boxy or bulky in shape, resulting in typical aerodynamic response.

Most buildings present closed, fairly smooth surfaces to the wind.

Most buildings are fit snugly to the ground, presenting a particular situation for the drag effects of the ground surface.

Most buildings have relatively stiff structures, resulting in a fairly limited range of variation of the natural period of vibration of the structure.

These and other considerations allow for the simplification of wind analysis by permitting a number of variables to be eliminated or to be lumped into a single modifying constant in the determination of wind effects. For unusual situations, such as elevated buildings, open structures, highly flexible structures, and unusual aerodynamic shapes, it may be advisable to do a more thorough analysis than the typical simplified one permitted by codes.

Based on the general assumptions previously enumerated, the total force of wind on a building is derived as an equivalent static pressure visualized in terms of a horizontally directed pressure on a vertical surface normal to the wind. The derivation of this pressure is in the form of the basic equation for kinetic energy and is expressed as

$$p = Cv^2$$

in which C is a combined adjustment for the air mass, the units used, and some of the assumptions previously described. With the wind in MPH and the pressure in lb/ft², the equation for buildings of rectangular form and moderate size becomes that used in the graph in Figure 3.1, which is

$$p = 0.003 \, v^2$$

This equation is not intended to represent the true pressure on any surface but is the net force resulting from the combination of direct pressure, suction, and drag. For use in determining the total force on the building, it is applied as a direct horizontal pressure on a vertical plane described by the silhouette of the building perpendicular to the direction of the wind.

Figure 4 in Chapter 23 of the *UBC* (see the Appendix) shows a map of the United States zoned with what is called the *minimum resultant wind pressure*. The map is intended for use as described in the preceding discussion. The pressures vary from 20 to 50 psf and may be seen on the graph in Figure 3.1 to relate to approximate wind velocities of 80 and 125 MPH respectively. An 80-MPH wind is quite violent and is seldom experienced in most parts of the United States. A wind of 125 MPH, on the other hand, is still below the maximum velocities estimated to be at the center of tornadoes and hurricanes (see Figure 3.1), which means that the code requirement actually provides safety only for the major peripheral winds of these storms.

As described on the *UBC* map, the minimum resultant wind pressures are those assumed to occur at the standard elevation of 30 ft above the ground. Table 23-F of the *UBC* provides for the conversion of this base pressure to the actual pressures on various height zones of buildings, accounting for the variation of the effects of ground drag.

The general effects of wind on stationary objects were described in the previous section. These effects are translated into building design criteria as explained in the following paragraphs.

Inward Pressure on Exterior Walls.   Surfaces directly facing the wind are usually required to be designed for the full base pressure, although this is somewhat conservative, because the windward force usually accounts for only about 60% of the total force on the building. Designing for only part of the total force is, however, partly compensated for by the fact that the base pressures are not generally related to gust effects, which tend to have less effect on the building as a whole and more effect on parts of the building.

Suction on Exterior Walls. Most codes also require suction on exterior walls to be the full base pressure, although the preceding comments about inward pressure apply here as well.

Pressure on Roof Surfaces. Depending on their actual form, as well as that of the building as a whole, nonvertical surfaces may be subjected to either inward or suction pressures because of wind. Actually such surfaces may experience both types of pressure as the wind shifts direction. Most codes require an uplift (suction) pressure equal to the full design pressure at the elevation of the roof level. Inward pressure is usually related to the actual angle of the surface as an inclination from the horizontal.

Overall Horizontal Force on the Building. Overall horizontal force is calculated as a horizontal pressure on the building silhouette, as previously described, with adjustments made for height above the ground. The lateral resistive structural system of the building is designed for this force.

Horizontal Sliding of the Building. In addition to the possible collapse of the lateral resistive system, there is the chance that the total horizontal force may slide the building off its foundations. For a tall building with fairly shallow foundations, this may also be a problem for the force transfer between the foundation and the ground. In both cases, the dead weight of the building generates a friction that helps to resist this force.

Overturn Effect. As with horizontal sliding, the dead weight tends to resist the overturn, or toppling, effect. In practice the overturn effect is usually analyzed in terms of the overturn of individual vertical elements of the lateral resistive system, rather than for the building as a whole.

Wind on Building Parts. The previously discussed clean-off effect is critical for elements that project from the general mass of the building. In some cases codes require for such elements a design pressure higher than the base pressure, so that gust effects as well as the clean-off problem are allowed for in the design.

Harmonic Effects.   Design for vibration, flutter, whipping, multi-nodal swaying, and so on requires a dynamic analysis and cannot be accounted for when using the equivalent static load method. Stiffening, bracing, and tightening of elements in general may minimize the possibilities for such effects, but only a true dynamic analysis or a wind tunnel test can assure the adequacy of the structure to withstand these harmonic effects.

Effect of Openings.   If the surface of a building is closed and reasonably smooth, the wind will slip around it in a fluid flow. Openings or building forms that tend to cup the wind can greatly affect the total wind force on the building. It is difficult to account for these effects in a mathematical analysis, except in a very empirical manner. Cupping of the wind can be a major effect when the entire side of a building is open, for example. Garages, hangars, band shells, and other buildings of similar form must be designed for an increased force that can only be estimated unless a wind tunnel test is performed.

Torsional Effect.   If a building is not symmetrical in terms of its wind silhouette, or if the lateral resistive system is not symmetrical within the building, the wind force may produce a twisting effect. This effect is the result of a misalignment of the centroid of the wind force and the centroid (called *center of stiffness*) of the lateral resistive system and will produce an added force on some of the elements of the structure.

Although there may be typical prevailing directions of wind in an area, the wind must be considered to be capable of blowing in any direction. Depending on the building shape and the arrangement of its structure, an analysis for wind from several possible directions may be required.

3.4   Building Code Requirements for Wind

Model building codes such as the *UBC* are technically not legal unless they are adopted by ordinances enacted by some state, county, or city. Although smaller communities usually adopt one of these model codes, states, counties, and large cities often develop their own codes, even though they may use one of the model codes as a basic reference. Where wind is a major local problem, local codes may add special requirements to those in the model code. The following is a discussion of the requirements for wind as taken from the 1979 edition of the *UBC*. Reprints of most of the material referred to are in the Appendix to this book.

Basic Design Pressure. Section 2311(a) of the *UBC* requires that the minimum design pressure be taken from Table 23–F of the *UBC*. The table shows various pressures for zones of the building at different distances above the ground. The basic reference zone is that from 30 to 49 ft, where the required design pressure is that recommended for the locality by the map in Figure 4 of the *UBC*. This is the basic design force that is applied as a horizontal pressure on the gross area of the vertical projection of the building—that is, on its profile or silhouette.

Uplift. Section 2311(c) requires that the roof of an enclosed building be designed for an uplift pressure of 75% of the required horizontal pressure at the elevation of the roof level. For unenclosed buildings and for overhangs, cornices, and so on, the uplift is required to be 125% of the horizontal pressure. This is a highly generalized provision, because various roof geometries can cause a wide range of actual aerodynamic effects. Experience has shown this provision to be reasonably conservative, although some unusual shapes can result in considerable uplift effect.

Sloping Roof Surfaces. Section 2311(d) of the *UBC* requires that sloping roof surfaces at an angle of over 30° be designed for an inward pressure normal to the surface and equal to the horizontal pressure at that elevation. This is also a highly generalized provision, with the pressure assumed to change suddenly from uplift to inward effect at 30°. Some codes give a more graduated set of values as the slope varies. Consideration must also be given to the possibility of the wind load in combination with the roof live load as the slope increases. This combination is discussed later. Reference should also be made to the variation of live load as given in *UBC* Section 2305(a) and Table 23–C.

The basic idea is that, if the wind blows at wind storm velocities, it is to be expected that any loose items on the roof (constituting the live load) are likely to be blown away before the critical wind velocity is reached. The principal exceptions are ice and wet snow, which adhere relatively tightly to the surface.

Solid Towers. Table 23–G of the *UBC* gives modifying factors for solid towers that consist of reductions in the total wind force when the tower cross section is rounded and has less aerodynamic drag.

Open Towers. Towers of open framework tend to let the wind slip through, reducing the total wind force from that considered to act on the overall profile of the tower. Section 2311(g) provides for the design of such structures using the actual profile of the parts with shape factors used to adjust both the total wind force on the tower and the loads on individual parts based on their shape.

Overturning Moment. Section 2311(i) requires that the overturning moment due to wind not exceed two-thirds of the dead load stabilizing moment, thus providing a theoretical safety factor of 1.5 against overturn. The usual design procedure is to determine a required overturn resisting moment of 1.5 times the actual wind effect. The dead load moment is then subtracted from the result, and if there is any excess overturn, it is provided for by anchorage of the structure against uplift. In most buildings the individual vertical elements of the lateral resistive system are anchored rather than the building as a whole.

Combined Wind and Live Load. Section 2311(j) requires that the required live load be included with the wind and dead load for stress calculations. An exception is given for snow load, for which it is required that 50% of the snow load be included. Because the code permits some reduction of the total live load when large areas are supported, there is some reduced probability already included when reference is made to the design load.

In various situations there may be many possible combinations

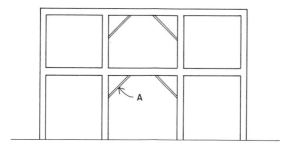

FIGURE 3.3   Knee-braced frame.

of loading to consider for different design purposes. In trusses and continuous frames, the critical consideration for an individual part may occur under partial, rather than full, loading. Overturn should be investigated without the live load.

The example in Figure 3.3 shows a two-story frame braced for lateral resistance by the use of knee braces in the middle bay. The load combinations to consider for the critical forces in the knee brace labeled "A" are the following:

1. Dead load only.
2. Dead load plus live load.
3. Dead load plus wind from the left.
4. Dead load plus live load plus wind from the left.
5. Dead load plus wind from the right.
6. Dead load plus live load plus wind from the right.

From this list the three potential critical combinations are 2, 3, and 6. Both 1 and 2 will produce compression in the knee brace, 2 being the higher load. Similarly, 5 and 6 will produce compression, 6 being the higher load. Although 6 will produce a higher force than 2, the one-third increase in allowable stress for wind may make the combination less critical; thus both cases must be investigated. Both 3 and 4 will produce potential tension in the knee brace if the wind effect exceeds the dead load and live load effects. Of the two, 3 is the more critical, because the wind has less opposition.

In structures with complex frameworks and considerable lack

of symmetry, this type of multiple case investigation can become quite laborious. Add the possibilities of moving loads or skip-stop live loading of adjacent bays, and the investigation can be a major undertaking unless computer-aided analysis is used. In most buildings, however, a careful study of the loading cases and their real potential critical nature will reduce the actual work to something reasonable.

Inward and Outward Pressure on Vertical Exterior Surfaces. The *UBC* does not specifically mention the problem of inward and outward pressure on vertical exterior surfaces, referring in Section 2311 only to the "minimum horizontal and uplift pressures." As was discussed previously in this chapter, the total force on the building has been empirically developed as the sum of the direct pressure, the drag, and the suction, in which case neither the direct pressure on the windward walls nor the suction on the leeward walls would be as great as the total wind force on the building. Some codes allow for this distribution effect by permitting the use of some reduced percentage for the pressure on individual surfaces. The usual assumption for rectangular buildings of average size is that the distribution is approximately 60% of the total force on the windward side and 40% on the leeward side. By omitting any reference to this assumption, the *UBC* implies that the direct and suction pressures should be the same as the overall wind force. Although the result is somewhat conservative, it is partly compensated for by gust effects, which tend to be more critical for smaller parts of the building than for the building as a whole.

Increased Pressure on Tall Buildings. Table 23–F of the *UBC* gives the wind pressures for various zones of distance above the ground. The figures account for the decreasing effect of ground surface drag as the height increases. Although the effect is actually a continuous one of variation with respect to height, and formulas exist for a smooth curve of variation, the usual code practice is to jump by zones as in the *UBC*.

Loads on Other Structures. Wind load requirements for miscellaneous structures other than buildings and for some parts of buildings are sometimes given in the codes. The requirements are often scattered through the code, instead of being included in the section on wind. Some of the *UBC* requirements of this type are:

Patio roof coverings: Section 4902 and Table 49–A.

Glazing: Section 5403 and Tables 54–A and 54–B.

Miscellaneous structures under 20 ft: Section 2311(h).

Loads on rooftop signs are sometimes separately specified, because they are highly susceptible to gust effects. The *UBC* does not make such provisions.

Various requirements and data for design for lateral loads are also to be found in other parts of the *UBC*. A general summary of these is included in the discussion of code requirements for seismic design in Section 4.4 of Chapter 4.

3.5 General Design Considerations for Wind

The relative importance of design for wind as an influence on the general building design varies greatly among buildings. The location of the building is a major consideration, the basic design pressure varying by a factor of 2.5 from the lowest wind area to the highest on the *UBC* map. Other important variations include the dead weight of the construction, the height of the building, the type of structural system (especially for lateral load resistance), the aerodynamic shape of the building and its exposed parts, and the existence of large openings, recessed portions of the surface, and so on.

The following is a discussion of some general considerations of design of buildings for wind effects. Any of these factors may be more or less critical in specific situations.

Influence of Dead Load. Dead load of the building construction is generally an advantage in wind design, because it is a stabilizing factor in resisting uplift, overturn, and sliding and tends to reduce the incidence of vibration and flutter. However, the stresses that result from various load combinations, all of which include dead load, may offset these gains when the dead load is excessive.

Anchorage for Uplift, Sliding, and Overturn. Ordinary connections between parts of the building may provide adequately for various transfers of wind force. In some cases, such as with lightweight elements, wind anchorage may be a major consideration. In most design cases the adequacy of ordinary construction details are considered first, and extraordinary measures are used only when required. Various situations of anchorage are illustrated in the examples in Chapters 6 and 7.

Critical Shape Considerations. Various aspects of the building form can cause increase or reduction in wind effects. Although it is seldom as critical in building design as it is for racing cars or aircraft, streamlining can improve the relative efficiency of the building in wind resistance. Some potential critical situations, as shown in Figure 3.4, are as follows:

1. Flat versus curved forms. Buildings with rounded forms, rather than rectangular forms with flat surfaces, offer less wind resistance.
2. Tall buildings that are short in horizontal dimension are more critical for overturn and possibly for the total horizontal deflection at their tops.
3. Open-sided buildings or buildings with forms that cup the wind tend to catch the wind, resulting in more wind force than that assumed for the general design pressures. Open structures must also be investigated for major outward force on internal surfaces.
4. Projections from the building. Tall parapets, solid railings, cantilevered balconies and parapets, wide overhangs, and free-standing exterior walls catch considerable wind and add to the overall drag effect on the building. Signs, chimneys, antennae, penthouses, and equipment on the roof of a building are also critical for the clean-off effect discussed previously.

Relative Stiffness of Structural Elements. In most buildings the lateral resistive structure consists of two basic elements: the horizontal distributing elements and the vertical cantilevered or braced frame elements. The manner in which the horizontal elements distribute forces and the manner in which the vertical

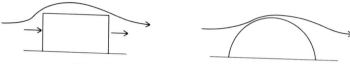

streamlining effect of rounded building forms

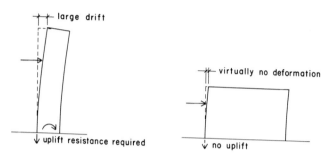

large drift

virtually no deformation

uplift resistance required

no uplift

overturn and drift related to the building profile

wind cupping effect of open sides and recesses

increased force on projecting elements

FIGURE 3.4 Wind effects related to building form.

elements share forces are critical considerations in wind analysis. The relative stiffness of individual elements is the major property that affects these relationships. The various situations that occur are discussed in Chapter 5 and illustrated in the examples in Chapters 6 and 7.

Stiffness of Nonstructural Elements. When the vertical elements of the lateral resistive system are relatively flexible, as with rigid frames and wood shear walls that are short in plan length, there may be considerable lateral force transferred to nonstructural elements of the building construction. Wall finishes of masonry veneer, plaster, or drywall can produce relatively rigid planes whose stiffnesses exceed those of the structures over which they are placed. If this is the case, the finish material may take the load initially, with the structure going to work only when the finish fails. This result is not entirely a matter of relative stiffness, however, because the load propagation through the building also depends on the attachments between elements of the construction. This problem should be considered carefully when developing the details of the building construction.

Allowance for Movement of the Structure. All structures deform when loaded. The actual dimension of movement may be insignificant, as in the case of a poured-concrete shear wall, or it may be considerable, as in the case of a slender steel rigid frame. The effect of these movements on other elements of the building construction must be considered. The case of transfer of load to nonstructural finish elements, as just discussed, is one example of this problem. Another critical example is that of windows and doors. Glazing must be installed so as to allow for some movement of the glass with respect to the frame. The frame must be installed so as to allow for some movement of the structure of the building, without load being transferred to the window frame.

All these considerations should be kept in mind in developing the general design of the building. If the building form and detail are determined and the choice of materials made before any thought is given to structural problems, it is not likely that an intelligent design will result. This is not to suggest that structural concerns are the most important concerns in building design but merely that they should not be relegated to afterthoughts.

4

Earthquake Effects
on Buildings

||||||||||||||||||||||||||||||||||||||||||||||||||||||||||||

Earthquakes are essentially vibrations of the earth's crust caused
by subterranean ground faults. They occur several times a day in
various parts of the world, although only a few each year are of
sufficient magnitude to cause significant damage to buildings.
Major earthquakes occur most frequently in particular areas of
the earth's surface that are called *zones of high probability*.
However, it is theoretically possible to have a major earthquake
anywhere on the earth at some time.

During an earthquake, the ground surface moves in all directions. The most damaging effects on structures are generally the
movements in a direction parallel to the ground surface (that is,
horizontally) because of the fact that structures are routinely
designed for vertical gravity loads. Thus, for design purposes the
major effect of an earthquake is usually considered in terms of
horizontal force, similar to the effect of wind.

A general study of earthquakes includes consideration of the
nature of ground faults, the propagation of shock waves through
the earth mass, the specific nature of recorded major quakes, and
so on. We do not present here a general discussion of earthquakes
but concentrate on their influence on the design of structures for

55

buildings. Some of the references listed after Chapter 8 may be used for a study of the general nature of earthquakes, and you are urged to study these if you lack such a background of knowledge.

4.1 Characteristics of Earthquakes

Following a major earthquake, it is usually possible to retrace its complete history through the recorded seismic shocks over an extended period. This period may cover several weeks, or even years, and the record will usually show several shocks preceding and following the major one. Some of the minor shocks may be of significant magnitude themselves, as well as being the foreshocks and aftershocks of the major quake.

A major earthquake is usually rather short in duration, often lasting only a few seconds and seldom more than a minute or so. During the general quake, there are usually one or more major peaks of magnitude of motion. These peaks represent the maximum effect of the quake. Although the intensity of the quake is measured in terms of the energy release at the location of the ground fault, called the *epicenter*, the critical effect on a given structure is determined by the ground movements at the location of the structure. The extent of these movements is affected mostly by the distance of the structure from the epicenter, but they are also influenced by the geological conditions directly beneath the structure and by the nature of the entire earth mass between the epicenter and the structure.

Modern recording equipment and practices provide us with representations of the ground movements at various locations, thus allowing us to simulate the effects of major earthquakes. Figure 4.1 shows the typical form of the graphic representation of one particular aspect of motion of the ground as recorded or as interpreted from the recordings for an earthquake. In this example the graph is plotted in terms of the acceleration of the ground in one horizontal direction as a function of elapsed time. For use in physical tests in laboratories or in computer modeling, records of actual quakes may be "played back" on structures in order to analyze their responses.

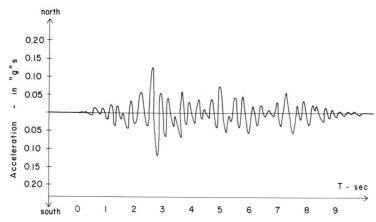

FIGURE 4.1   Characteristic form of ground acceleration graph for an earthquake.

These playbacks are used in research and in the design of some major structures to develop criteria for design of lateral resistive systems. Most building design work, however, is done with criteria and procedures that have been evolved through a combination of practical experience, theoretical studies, and some empirical relationships derived from research and testing. The results of the current collective knowledge are put forth in the form of recommended design procedures and criteria that are incorporated into the building codes.

Although it may seem like a gruesome way to achieve it, we advance our level of competency in design every time there is a major earthquake that results in some major structural damage to buildings. Engineering societies and other groups routinely send investigating teams to the sites of major quakes to report on the effects on buildings in the area. Of particular interest are the effects on recently built structures, because these buildings are, in effect, full-scale tests of the validity of our most recent design techniques. Each new edition of the building codes usually reflects some of the results of this cumulative growth of knowledge culled from the latest disasters.

4.2   General Effects of Earthquakes

The ground movements caused by earthquakes can have several types of damaging effects. Some of the major effects are:

Direct movement of structures. Direct movement is the motion of the structure caused by its attachment to the ground. The two primary effects of this motion are a general destabilizing effect due to the shaking and to the impelling force caused by the inertia of the structure's mass.

Ground surface faults. Surface faults may consist of cracks, vertical shifts, general settlement of an area, landslides, and so on.

Tidal waves. The ground movements can set up large waves on the surface of bodies of water that can cause major damage to shoreline areas.

Flooding, fires, gas explosions, and so on. Ground faults or movements may cause damage to dams, reservoirs, river banks, buried pipelines, and so on, which may result in various forms of disaster.

Although all these possible effects are of concern, we deal in this book only with the first effect: the direct motion of structures. Concern for this effect motivates us to provide for some degree of dynamic stability (general resistance to shaking) and some quantified resistance to energy loading of the structure.

The force effect caused by motion is generally directly proportional to the dead weight of the structure—or more precisely, to the dead weight borne by the structure. This weight also partly determines the character of dynamic response of the structure. The other major influences on the structure's response are its fundamental period of vibration and its efficiency in energy absorption. The vibration period is basically determined by the mass, the stiffness, and the size of the structure. Energy efficiency is determined by the elasticity of the structure and by various factors such as the stiffness of supports, the number of independently moving parts, the rigidity of connections, and so on.

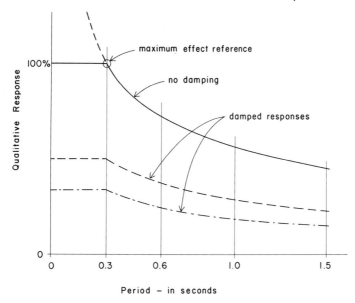

FIGURE 4.2   Spectrum response graph.

A relationship of major concern is that which occurs between the period of the structure and that of the earthquake. Figure 4.2 shows a set of curves, called *spectrum curves*, that represent this relationship as derived from a large number of earthquake "playbacks" on structures with different periods. The upper curve represents the major effect on a structure with no damping. Damping results in a lowering of the magnitude of the effects, but a general adherence to the basic form of the response remains.

The general interpretation of the spectrum effect is that the earthquake has its major direct force effect on buildings with short periods. These tend to be buildings with stiff lateral resistive systems, such as shear walls and X-braced frames, and buildings that are small in size and/or squat in profile.

For very large, flexible structures, such as tall towers and high-rise buildings, the fundamental period may be so long that the structure develops a whiplash effect, with different parts of the structure moving in opposite directions at the same time, as

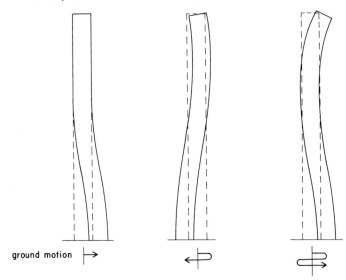

ground motion

FIGURE 4.3 Earthquake motion of a tall building.

shown in Figure 4.3. Analysis for this behavior requires the use of dynamic methods that are beyond the scope of this book. The three general cases of structural response are illustrated by the three cases shown in Figure 4.4. Referring to the spectrum curves, for buildings with a period below that representing the upper cutoff of the curves (approximately 0.3 sec), the response is that of a rigid structure, with virtually no flexing. For buildings with a period slightly higher, there is some reduction in the force effect caused by the slight "giving" of the building and its using up some of the energy of the motion-induced force in its own motion. As the building period increases, the behavior approaches that of the slender tower, as shown in Figure 4.3.

In addition to the movement of the structure as a whole, there are independent movements of individual parts. These each have their own periods of vibration, and the total motion occurring in the structure can thus be quite complex if it is composed of a number of relatively flexible parts.

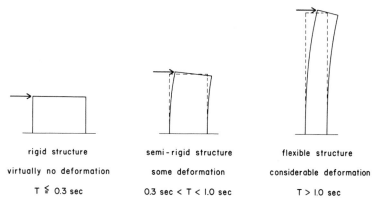

rigid structure   semi-rigid structure   flexible structure

virtually no deformation   some deformation   considerable deformation

T ≲ 0.3 sec   0.3 sec < T < 1.0 sec   T > 1.0 sec

FIGURE 4.4   Seismic response of buildings.

4.3   Earthquake Effects on Buildings

The principal concern in structural design for earthquake forces is for the laterally resistive system of the building. In most buildings this system consists of some combination of horizontally distributing elements (usually roof and floor diaphragms) and vertical bracing elements (shear walls, rigid frames, trussed bents, etc.). Failure of any part of this system, or of connections between the parts, can result in major damage to the building, including the possibility of total collapse.

It is well to remember, however, that an earthquake shakes the whole building. If the building is to remain completely intact, the potential movement of all its parts must be considered. The survival of the structural system is a limited accomplishment if suspended ceilings fall, windows shatter, plumbing pipes burst, and elevators are derailed.

A major design consideration is that of tying the building together so that it is quite literally not shaken apart. With regard to the structure, this means that the various separate elements must be positively secured to one another. The detailing of construction connections is a major part of the structural design for earthquake resistance.

In some cases it is desirable to allow for some degree of independent motion of parts of the building. This is especially critical in situations where a secure attachment between the structure and various nonstructural elements, such as window glazing, can result in undesired transfer of force to the nonstructural elements. In these cases use must be made of connecting materials and details that allow for the holding of the elements in place while still permitting relative independence of motion.

When the building form is complex, various parts of the building may tend to move differently, which can produce critical stresses at the points of connection between the parts of the building. The best solution to this sometimes is to provide connections (or actually in some cases nonconnections) that allow for some degree of independent movement of the parts. This type of connection is called a *seismic separation joint*, and its various problems are discussed in Section 8.2 of Chapter 8.

Except for the calculation and distribution of the loads, the design for lateral loads from earthquakes is generally similar to that for the horizontal forces that result from wind. In some cases the code requirements are the same for the two loading conditions. There are many special requirements for seismic design in the *UBC*, however, and the discussion in the next section, together with the examples in Chapter 7, deal with the use of the code for analysis and design for earthquake effects.

4.4   Building Code Requirements for Earthquakes

The following is a discussion of the various requirements of the 1979 edition of the *UBC* for seismic design. The code for seismic design is usually quite up-to-date, a new edition being published every three years. The main body of material on seismic design is in Section 2312. We begin with a discussion of this section and proceed to a discussion of some materials to be found in other sections of the code. See the Appendix for reprints of the code.

Section 2312: General Design Requirements.   In the beginning of this section are definitions of various terms and symbols used in the code. Reference should be made to this portion whenever any term or symbol is not clear to the reader.

The basic formula for determination of the lateral load on the structure is *UBC* formula 12–1:

$$V = Z\,I\,K\,C\,S\,W$$

in which:

Z varies from 3/16 to 1, depending on the seismic zone as identified in the maps in *UBC* Figures 1, 2, and 3.

I is a factor relating to the occupancy or use of the building, as given in *UBC* Table 23–K.

K is based on the type of lateral resistive structural system and is given in *UBC* Table 23–I.

C is a factor empirically derived from the relationship of the building period to the average earthquake effect, as was discussed previously in terms of the spectrum curves.

S is a factor that accounts for the effect of local ground conditions, which may dampen or amplify the shock.

W is the dead weight of the building.

C is determined from *UBC* formula 12–2 as

$$C = \frac{1}{15\sqrt{T}}$$

in which *T* is the fundamental period of the building in seconds.

In the words of the code, *T* is to be established by a "properly substantiated analysis." A recommended formula for this determination is given in *UBC* formula 12–3, which incorporates considerations of the seismic load distribution and the deformation characteristics of the structure. This is a somewhat complex analysis, so the code permits alternative determinations by the simpler formulas, 12–3A and 12–3B.

Formula 12–3A expresses *T* as a function of the height of the structure and its depth in the direction of the lateral load.

$$T = \frac{0.05\,h_n}{\sqrt{D}}$$

Formula 12–3B expresses T simply as a function of the number of stories of the structure.

$$T = 0.10\,N$$

Formula 12–3B is to be used only for buildings in which the lateral resistive structure consists of a ductile rigid frame.

The *UBC* stipulates that the value of C need not exceed 0.12. Entering this value in the formula for C, we can derive a lower limit for T. Thus

If $$C = \frac{1}{15\,\sqrt{T}} = 0.12$$

Then $$T = \left[\frac{1}{15(0.12)}\right]^2 = 0.309 \text{ sec}$$

Figure 4.5 shows a plot of C as a function of T, using *UBC* formula 12–2. The form may be seen to be that of the typical spectrum curve. The maximum value for C is seen to correspond to the derived value of 0.309 sec for T. Below the graph are shown some interpretations of *UBC* formulas 12–3A and 12–3B that relate to the values for T in the graph.

Calculation of the S factor requires a determination of the fundamental period of the site (T_s) as well as of the structure. Determining T_s requires rather extensive geological information of the site conditions. In the absence of this information, the *UBC* permits the use of a maximum value for the product of C and S of 0.14. Because the code also requires a minimum value for S of 1.0, if the maximum value of C of 0.12 is used, the bracketed range for the product of C and S is between 0.12 and 0.14. The net result of this is that, unless the building project cost is high, the true C value is less than 0.12, or if the necessary geological information is available from previous studies in the immediate vicinity, the maximum CS value of 0.14 is used for most designs.

The use of the S factor is a fairly recent addition to the code, and there is some controversy regarding its validity as a requirement for all structures. In time the use of this factor may be more rationally developed in the code criteria so as not to amount to a de facto general increase in the required lateral load.

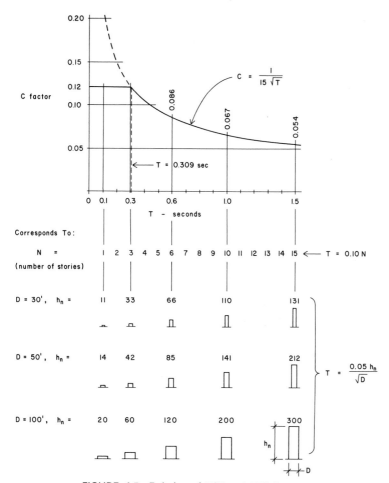

FIGURE 4.5 Relation of "C" and "T" factors.

65

Subsection (e) of Section 2312 deals with the distribution of seismic loads on the structure. Formula 12–5 requires that the total load on the structure be

$$V = F_t + \sum_{i=1}^{n} F_i$$

The summation portion of this formula consists of the loads as normally calculated and applied at the various levels of the structure. F_t is an added force that must be applied at the top of the structure as well as being included in the total, and it is found using formula 12–6.

$$F_t = 0.07 \, TV$$

The maximum required value for F_t is 0.25 V and F_t may be considered to be zero where T is 0.7 sec or less. As shown in Figure 4.5, a T value this high generally occurs only with relatively tall buildings.

The value of the lateral shear force at each level of a multilevel structure (F_x) is determined from formula 12–7. The use of this formula for the analysis of multistory structures is illustrated in the examples in Chapter 7.

The code formulas are generally applicable only to buildings that are reasonably simple in shape and have little change in their basic structural systems or their plans from top to bottom. Changes in these characteristics can result in behaviors more complex than those assumed for the general formulas and their limits. Subsection 2312(e)3 of the *UBC* requires that these more complex structures be analyzed by dynamic methods, rather than by the simpler equivalent static load method. An exception is made in Subsection 2312(e)2 for setbacks of less than 25% in multistory buildings.

Subsection 2312(e)5 requires that the structure be analyzed for torsion when the center of rigidity of the structure does not coincide with the centroid of the load. Even when this eccentricity is not present, the code requires design for a minimum assumed eccentricity of 5% of the maximum plan dimension of the building when horizontal diaphragms are used to distribute the shear.

Overturning of the building or the structure is provided for in Subsection 2312(f). When the lateral resistive system has a number of isolated elements (such as individual shear walls), the overturn effect is distributed to them in the same manner as for the distribution of the lateral shear forces.

Subsection 2312(g) provides that parts of the structure be designed for individual lateral forces that result from the parts' own weight. This force is calculated from formula 12–8 as

$$F_p = Z\,I\,C_p\,W_p$$

The C_p factor in this equation replaces the C, S, and K factors as used in the total load for the building in formula 12–1. Values for C_p are given in *UBC* Table 23–J for various types of elements and situations.

Subsection 2312(h) limits the relative deflection of one story with respect to another (called the *drift*) to 0.5% of the story height, unless an analysis can demonstrate that more deflection is tolerable. A multiplying factor of $1/K$ must be used for the deflection calculation when K is less than 1 for the structure.

Subsection 2312(j) contains a number of requirements for the design of specific structural systems and elements. Reference is made to these requirements in the examples in Chapters 6 and 7.

Subsection 2312(k) defines essential facilities as they are listed in Table 23–K for establishment of the I factor in the lateral load formula, 12–1.

Lateral Load Requirements in Other Chapters. The following is a summary of some of the requirements for lateral load design to be found in other portions of the *UBC*. In some cases these requirements apply to both wind and seismic loads.

1. Stress increase: Section 2303(d). This section contains the provision for a one-third increase in allowable stresses when load combinations include wind or seismic force.
2. Load combinations: Section 2303(f). The highly remote possibility of a simultaneous earthquake and wind storm loading is excluded by this section.
3. Anchorage of masonry and concrete walls: Section 2310. This section requires that the connections of masonry and concrete walls to roofs and floors that provide lateral

support for the walls be designed for a minimum load of 200 lb/ft.

4. Increase of force on masonry shear walls: Table 24–H. The third footnote to this table requires that the seismic force on reinforced masonry shear walls be increased by 50% over that required by Section 2312 when investigating shear stress in the wall.

5. Wood diaphragms: Section 2514. This is an extensive section with data and requirements for plywood and board-sheathed diaphragms.

6. Fiberboard sheathed diaphragms: Section 2515. This is a special section with design requirements for vertical diaphragms of fiberboard on a wood frame. The fiberboard referred to must meet the requirements of one of the *UBC* standards. Allowable loads for these diaphragms are given in Table 25–0.

7. Bracing of stud walls: Section 2518(g)5. This section requires that every exterior wall and main interior wall be braced at or near its ends and at least every 25 ft throughout its length. The approved methods of bracing include diagonal, let-in braces, diagonal board sheathing, plywood sheathing, fiberboard sheathing, gypsum sheathing, particleboard sheathing, gypsum drywall, and cement plaster. Reference is made to various code tables for the load capacities of these various constructions. Although all these wall constructions produce potential shear walls, the details for anchorage and load transfer must be developed in order to use them.

8. Maximum diaphragm dimension ratios for wood diaphragms: Table 25–I. This table gives the limits for the span-to-width ratios of simple-span horizontal diaphragms and the height-to-width ratios for vertical diaphragms with plywood or diagonal board sheathing.

9. Allowable loads on horizontal plywood diaphragms: Table 25–J. This table gives the allowable shear load per foot of width for horizontal plywood diaphragms. Variables include the type and thickness of plywood, size and spacing of nails, and details of the framing and panel layout.

10. Allowable loads on vertical plywood diaphragms: Table 25–K. This table is similar to Table 25–J, giving the same information for shear walls with plywood sheathing.

11. Minimum nailing requirements: Table 25–P. This table gives minimum nailing requirements for the various elements of light wood-frame construction. Reference should be made to this table when developing the details for structural connections to assure that calculated nailing is not used when it falls below the requirements of this table.

12. Concrete rigid frames: Sections 2625 and 2626. These sections give the requirements for ductile frames of reinforced concrete.

13. Concrete shear walls and trussed frames: Section 2627. This section gives requirements for concrete shear walls and "braced frames" (trussed frames, as defined by *UBC* 2312.)

14. Steel rigid frames: Sections 2722 and 2723. These sections give the requirements for steel rigid frames used to resist seismic load.

15. Wood-framed walls with miscellaneous sheathing: Section 4713 and Tables 47–H and 47–I. This material gives requirements and data for wood-framed shear walls faced with various materials other than plywood, diagonal boards, or fiberboard.

Although the *UBC* is reasonably well organized and indexed, it takes some time to gain a familiarity with all the material that is pertinent to structural design for wind and earthquakes. The use of much of the material just described is illustrated in the examples in Chapters 6 and 7 of this text, but not all possible situations can be covered in a limited number of examples.

4.5 General Design Considerations for Earthquake Forces

The influence of earthquake considerations on the design of building structures tends to be the greatest in the zones of highest probability of quakes. This fact is directly reflected in the *UBC* by the Z factor, which varies from 3/16 to 1, or by a ratio of more than 5 to 1. As a result, wind factors often dominate the design in the zones of lower probability.

A number of general considerations in the design of lateral resistive systems were discussed at the end of Chapter 3. Most of these also apply to seismic design. Some additional considerations are discussed in the following paragraphs.

Influence of Dead Load.   Dead load is in general a disadvantage in earthquakes, because the lateral force is directly proportional to it. Care should be exercised in developing the construction details and in choosing materials for the building in order to avoid creating unnecessary dead load, especially at upper levels in the building. Dead load is useful for overturn resistance and is a necessity for the foundations that must anchor the building.

Advantage of Simple Form and Symmetry.   Buildings with relatively simple forms and with some degree of symmetry usually have the lowest requirements for elaborate or extensive bracing or for complex connections for lateral loads. Design of plan layouts and of the building form in general should be done with a clear understanding of the ramifications in terms of structural requirements when wind or seismic forces are high. When complex form is deemed necessary, the structural cost must be acknowledged.

Following Through with Load Transfers.   It is critical in design for lateral loads that the force paths be complete. Forces must travel from their points of origin through the whole system and into the ground. Design of the connections between elements and of the necessary drag struts, collectors, chords, blocking, hold downs, and so on is highly important to the integrity of the whole lateral resistive system. The ability to visualize the load paths and a reasonable understanding of building construction details and processes are prerequisites to this design work.

Use of Positive Connections.   Earthquake forces often represent the most severe demands on connections because of their dynamic, shaking effects. Many means of connection that may be adequate for static force resistance fail under the jarring, loosening effects of earthquakes. Failures of a number of recently built buildings and other structures in earthquakes have been due to connection failures, even though the structures were designed in accordance with current code requirements and accepted practices. Increasing attention is being paid to this problem in the development of recommended details for building construction.

5

Elements of Lateral Load Resistive Systems

||||||||||||||||||||||||||||||||||||||||||||||||||||||||||||

The purpose of this chapter is to present a general discussion of some of the basic components of lateral load resistive systems. The analysis and design of specific cases is illustrated in the examples in Chapters 6 and 7. The intention in this chapter is to present the fundamental behavior and general problems of these elements so that the specific examples may be somewhat more briefly developed. The elements discussed here include the following:

Horizontal diaphragms.
Vertical diaphragms, or shear walls.
Braced frames—trussed or X-braced.
Rigid frames, or moment-resistive frames.

Also included is a discussion of some of the special devices and elements required for the use of these system components and for their interconnection and anchorage. This discussion includes items such as tie downs, drag struts, diaphragm chords, collectors, and framing anchors.

5.1 Horizontal Diaphragms

Most lateral resistive structural systems for buildings consist of some combination of vertical elements and horizontal elements. The most common vertical elements are shear walls, braced (trussed) frames, and rigid frames. The horizontal elements are most often the roof and floor framing and the deck systems. When the latter contain decks of sufficient strength and stiffness to be developed as rigid planes, they are called *horizontal diaphragms*. Decks most commonly used for this purpose are those constructed of plywood, diagonally placed wood boards, formed sheet steel, and poured concrete.

A horizontal diaphragm typically functions by collecting the lateral forces at a particular level of the building and then distributing them to the vertical elements of the lateral resistive system. For wind forces, the lateral loading of the horizontal diaphragm is usually through the attachment of the exterior walls to its edges. For seismic forces, the loading is partly a result of the weight of the deck itself and partly a result of the weights of other parts of the building that are attached to it.

The particular structural behavior of the horizontal diaphragm and the manner in which loads are distributed to vertical elements depend on a number of considerations that are best illustrated by various example cases in Chapters 6 and 7. Some of the general issues of concern are discussed in the following paragraphs.

Relative Stiffness of the Horizontal Diaphragm.   If the horizontal diaphragm is relatively flexible, it may deflect so much that its continuity is negligible and the distribution of load to the relatively stiff vertical elements is essentially on a peripheral basis. If the deck is quite rigid, on the other hand, the distribution to vertical elements will be essentially in proportion to their relative stiffness with respect to each other. The possibility of these two situations is illustrated for a simple box system in Figure 5.1.

Torsional Effects.   If the centroid of the lateral forces in the horizontal diaphragm does not coincide with the centroid of the stiffness of the vertical elements, there will be a twisting action

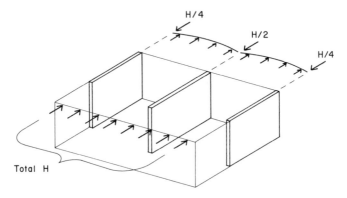

Peripheral distribution – flexible horizontal diaphragm

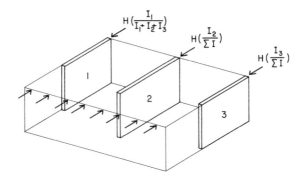

Proportionate stiffness distribution – rigid horizontal diaphragm

FIGURE 5.1   Distribution of load from a horizontal diaphragm.

(called *rotation effect* or *torsional effect*) on the structure as well as the direct force effect. Figure 5.2 shows a structure in which this effect occurs because of a lack of symmetry of the structure. This effect is usually of significance only if the horizontal diaphragm is relatively stiff. This stiffness is a matter of the materials of the construction as well as the depth-to-span ratio of the horizontal diaphragm. In general, wood and metal decks are quite flexible, whereas poured concrete decks are very stiff. In the example in Figure 5.2, even with a wood or metal deck, the

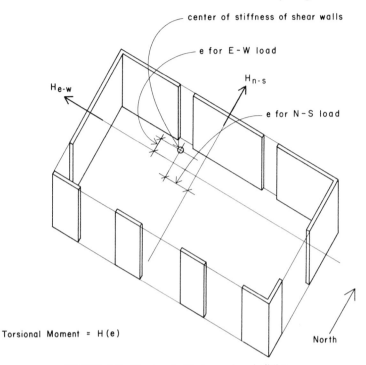

FIGURE 5.2 Torsional effect on a box building.

torsional effect would probably be considered in the east–west direction but not in the north–south direction, because of the difference of the proportions of the depth and span in the two directions.

Relative Stiffness of the Vertical Elements. When vertical elements share load from a rigid horizontal diaphragm, as shown in the lower figure in Figure 5.1, their relative stiffness must usually be determined in order to establish the manner of the sharing. The determination is comparatively simple when the elements are similar in type and materials, such as all-plywood shear walls. When the vertical elements are different, such as a mix of plywood and masonry shear walls or of some shear walls and some braced frames, their actual deflections must be calculated in order to establish the distribution, and this may require laborious calculations. A general discussion of deflections due to lateral loads is given in Section 8.1 of Chapter 8.

Use of Control Joints   The general approach in design for lateral loads is to tie the whole structure together to assure its overall continuity of movement. Sometimes, however, because of the irregular form or large size of a building, it may be desirable to control its behavior under lateral loads by the use of structural separation joints. In some cases these joints function to create total separation, allowing for completely independent motion of the separate parts of the building. In other cases the joints may control movements in a single direction while achieving connection for load transfer in other directions. A general discussion of separation joints is given in Section 8.2 of Chapter 8.

In performing their basic tasks, horizontal diaphragms have a number of potential stress problems. A major consideration is that of the shear stress in the plane of the diaphragm, caused by the spanning action of the diaphragm as shown in Figure 5.3. This spanning action results in shear stress in the material as well as a force that must be transferred across joints in the deck when the deck is composed of separate elements such as sheets of plywood or units of formed sheet metal. The sketch in Figure 5.4 shows a typical plywood framing detail at the joint between two sheets. The stress in the deck at this location must be passed from one sheet through the edge nails to the framing member and then back out through the other nails to the adjacent sheet.

As is the usual case with shear stress, both diagonal tension and diagonal compression are induced simultaneously with the shear stress. The diagonal tension becomes critical in materials such as concrete. The diagonal compression is a potential source of buckling in decks composed of thin sheets of plywood or metal. In plywood decks the thickness of the plywood relative to the spacing of framing members must be considered, and it is also a reason why the plywood must be nailed to intermediate framing members (not at edges of the sheets) as well as at edges. In metal decks the gauge of the sheet metal and the spacing of stiffening ribs must be considered. Tables of allowable loads for various

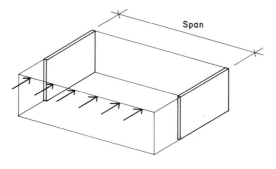

Span

Beam Analogy

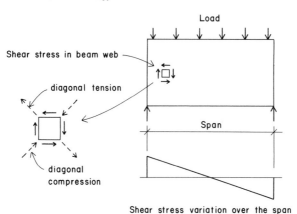

Load

Shear stress in beam web

diagonal tension

diagonal
compression

Span

Shear stress variation over the span

FIGURE 5.3   Functions of a horizontal diaphragm.

deck elements usually incorporate some limits for these considerations.

Diaphragms with continuous deck surfaces are usually designed in a manner similar to that for webbed steel beams. The
web (deck) is designed for the shear, and the flanges (edge-
framing elements) are designed to take the moment, as shown in
Figure 5.5. The edge members are called *chords*, and they must
be designed for the tension and compression forces at the edges.
With diaphragm edges of some length, the latter function usually

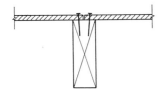

FIGURE 5.4   Nailed joint at plywood panel edge.

requires that the edge members be spliced for some continuity of the forces. In many cases there are ordinary elements of the framing system, such as spandrel beams or top plates of stud walls, that have the potential to function as chords for the diaphragm.

In some cases the collection of forces into the diaphragm or the distribution of loads to vertical elements may induce a stress beyond the capacity of the deck alone. Figure 5.6 shows a building in which a continuous roof diaphragm is connected to a series of shear walls. Load collection and force transfers require that some force be dragged along the dotted lines shown in the figure. For the outside walls it is possible that the edge framing used for chords can do double service for this purpose. For the interior shear wall, and possibly for the edges if the roof is cantilevered past the walls, some other framing elements may be necessary to reinforce the deck.

The diaphragm shear capacities for commonly used decks of various materials are available from the codes or from load tables prepared by deck manufacturers. Loads for plywood decks are

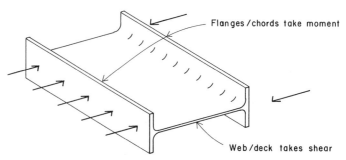

FIGURE 5.5   Horizontal diaphragm functions—beam analogy.

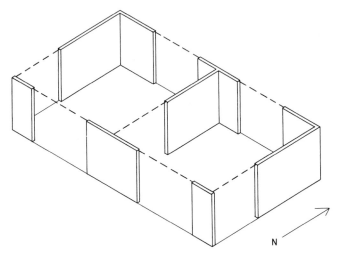

FIGURE 5.6   Collector functions in a box building.

given in the *UBC* (see the Appendix). An example of a lateral load table for steel deck is also given in the Appendix. Other tabulations are available from product manufacturers, although care should be exercised in their use to be certain that they are acceptable to the building code of jurisdiction.

A special situation is that of a horizontal system that consists partly or wholly of a braced frame. Care may be required when there are a large number of openings in the roof deck, or when the diaphragm shear stress is simply beyond the capacity of the deck. In the event of a deck with no code-accepted rating for shear, the braced frame may have to be used for the entire horizontal system. This and other types of braced frames are discussed in Section 5.3.

The horizontal deflection of flexible decks, especially those with high span-to-depth ratios, may be a critical factor in their design. Calculation of actual deflection dimensions may be required to determine the effect on vertical elements of the building construction or to establish positively whether the deck must be considered as essentially flexible or rigid, as discussed previously.

The use of subdiaphragms may also be required in some cases, necessitating the design of part of the whole system as a separate diaphragm, even though the deck may be continuous.

5.2 Vertical Diaphragms

Vertical diaphragms are usually the walls of buildings. As such, in addition to their shear wall function, they must fulfill various architectural functions and may also be required to serve as bearing walls for the gravity loads. The location of walls, the materials used, and some of the details of their construction must be developed with all these functions in mind.

The most common shear wall constructions are those of poured concrete, masonry, and wood frames of studs with some surfacing elements. Wood frames may be made rigid in the wall plane by the use of diagonal bracing or by the use of surfacing materials that have sufficient strength and stiffness. Choice of the type of construction may be limited by the magnitude of shear caused by the lateral loads but will also be influenced by fire code requirements and the satisfaction of the various other wall functions, as described previously.

Some of the structural functions usually required of vertical diaphragms are the following (see Figure 5.7):

1. *Direct shear resistance* usually consists of the transfer of a lateral force in the plane of the wall from some upper level of the wall to a lower level or to the bottom of the wall. This results in the typical situation of shear stress and the accompanying diagonal tension and compression stresses, as discussed for horizontal diaphragms.
2. *Cantilever moment resistance.* Shear walls generally work like vertical cantilevers, developing compression on one edge and tension on the opposite edge, and transferring an overturning moment (M) to the base of the wall.
3. *Horizontal sliding resistance.* The direct transfer of the lateral load at the base of the wall produces the tendency for the wall to slip horizontally off of its supports.

The shear stress function is usually considered independently of other structural functions of the wall. The maximum shear stress

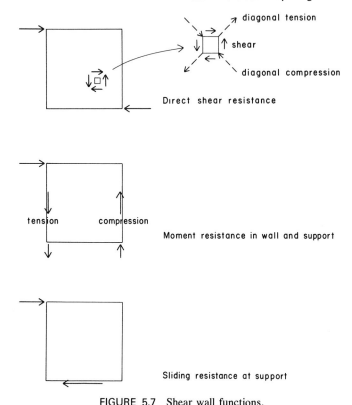

diagonal tension

shear

diagonal compression

Direct shear resistance

tension   compression

Moment resistance in wall and support

Sliding resistance at support

FIGURE 5.7   Shear wall functions.

that derives from lateral loads is compared to some rated capacity of the wall construction, with the usual increase of one-third in allowable stresses because the lateral load is most often a result of wind or earthquake forces. For concrete and masonry walls, the actual stress in the material is calculated and compared with the allowable stress for the material. For structurally surfaced wood frames, the construction as a whole is generally rated for its total resistance in pounds per ft of the wall length in plan. For a plywood surfaced wall, this capacity depends on the type and thickness of the plywood; the size, wood species, and spacing of the studs; the size and spacing of the plywood nails; and the inclusion or omission of blocking at horizontal plywood joints.

The analysis and design of shear walls of concrete and reinforced masonry are illustrated in the examples in Chapters 6 and 7. For walls of concrete, the procedures are based on the requirements of the American Concrete Institute (ACI) code (Ref. 6). For walls of reinforced hollow concrete block construction the procedures are based on the requirements of Chapter 24 of the *UBC*.

For wood stud walls the *UBC* provides tables of rated load capacities for several types of surfacing, including plywood, diagonal wood boards, plaster, gypsum drywall, and fiberboard. This material is included in the *UBC* reprints in the Appendix.

Although the possibility exists for the buckling of walls as a result of the diagonal compression effect, this is usually not critical because other limitations exist to constrain wall slenderness. The thickness of masonry walls is limited by maximum values for the ratio of unsupported wall height or length-to-wall thickness. Concrete thickness is usually limited by forming and pouring considerations, so that thin walls are not common except with precast construction. Slenderness of wood studs is limited by gravity design and by the code limits as a function of the stud size. Because stud walls are usually surfaced on both sides, the resulting sandwich-panel effect is usually sufficient to provide a reasonable stiffness.

As in the case of horizontal diaphragms, the moment effect on the wall is usually considered to be resisted by the two vertical edges of the wall acting as flanges or chords. In the concrete or masonry wall, this results in a consideration of the ends of the wall as columns, sometimes actually produced as such by thickening of the wall at the ends. In wood-framed walls, the end framing members are considered to fulfill this function. These edge members must be investigated for possible critical combinations of loading because of gravity and the lateral effects.

The overturn effect of the lateral loads must be resisted with the safety factor of 1.5 that is required by the *UBC*. The form of the analysis for the overturn effect is as shown in Figure 5.8. If the tiedown force is actually required, it is developed by the anchorage of the edge-framing elements of the wall.

Resistance to horizontal sliding at the base of a shear wall is

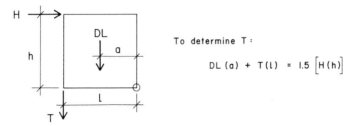

To determine T :

$$DL\,(a) + T\,(l) = 1.5\,\Big[H\,(h)\Big]$$

FIGURE 5.8   Overturn analysis for a shear wall.

usually at least partly resisted by friction caused by the dead loads. For masonry and concrete walls with dead loads that are usually quite high, the frictional resistance may be more than sufficient. If it is not, shear keys must be provided. For wood-framed walls, the friction is usually ignored and the sill bolts are designed for the entire lateral load.

An important judgment that must often be made in designing for lateral loads is that of the manner of distribution of lateral force between a number of shear walls that share the load from a single horizontal diaphragm. In some cases the existence of symmetry or of a flexible horizontal diaphragm may simplify this consideration. In many cases, however, the relative stiffnesses of the walls must be determined for this calculation.

If considered in terms of static force and elastic stress/strain conditions, the relative stiffness of a wall is inversely proportionate to its deflection under a unit load. Figure 5.9 shows the manner of deflection of a shear wall for two assumed conditions. In (a) the wall is considered to be fixed at its top and bottom, flexing in a double curve with an inflection point at midheight. This is the case usually assumed for a continuous wall of concrete or masonry in which a series of individual wall portions (called *piers*) are connected by a continuous upper wall or other structure of considerable stiffness. In (b) the wall is considered to be fixed at its bottom only, functioning as a vertical cantilever. This is the case for independent, free-standing walls or for walls in which the continuous upper structure is relatively flexible. A third possibility is shown in (c), in which relatively short piers are assumed to

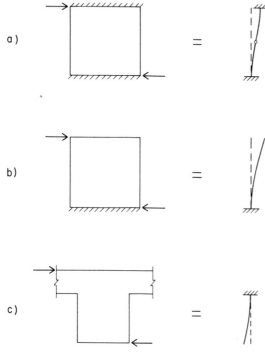

FIGURE 5.9   Shear wall support conditions: (a) fixed top and bottom, (b), and (c) cantilevered.

be fixed at their tops only, which produces the same deflection condition as in (b).

In some cases the deflection of the wall may result largely from shear distortion, rather than from flexural distortion, perhaps because of the wall materials and construction or the proportion of wall height to plan length. Furthermore, stiffness in resistance to dynamic loads is not quite the same as stiffness in resistance to static loads. The following recommendations are made for single-story shear walls:

1.   For wood-framed walls with height-to-length ratios of 2 or less, assume the stiffness to be proportional to the plan length of the wall.

2. For wood-framed walls with height-to-length ratios over 2 and for concrete and masonry walls, assume the stiffness to be a function of the height-to-length ratio and the method of support (cantilevered or fixed top and bottom). Use the values for pier rigidity given in the tables in the *Concrete Masonry Design Manual* (Ref. 7; see also the Appendix).
3. Avoid situations in which walls of significantly great differences in stiffness share loads along a single row. The short walls will tend to receive a small share of the loads, especially if the stiffness is assumed to be a function of the height-to-length ratio.
4. Avoid mixing of shear walls of different construction when they share loads on a deflection basis.

Item 4 in the preceding list can be illustrated by two situations as shown in Figure 5.10. The first situation is that of a series of panels in a single row. If some of these panels are of concrete or masonry and others of wood-frame construction, the stiffer concrete or masonry panels will tend to absorb the major portion of the load. The load sharing must be determined on the basis of actual calculated deflections. Better yet is a true dynamic analysis because if the load is truly dynamic in character, the periods of the two types of walls are of more significance than their stiffness.

In the second situation shown in Figure 5.10, the walls share load from a rigid horizontal diaphragm. This situation also requires a deflection calculation in order to determine the distribution of force to the panels.

In addition to the various considerations mentioned for the shear walls themselves, care must be taken to assure that they are properly anchored to the horizontal diaphragms. Problems of this sort are illustrated in the examples in Chapters 6 and 7.

A final consideration for shear walls is that they must be made an integral part of the whole building construction. In long building walls with large door or window openings or other gaps in the wall, shear walls are often considered as entities (isolated, independent piers) for their design. However, the behavior of the entire wall under lateral load should be studied to be sure that elements not considered to be parts of the lateral resistive system

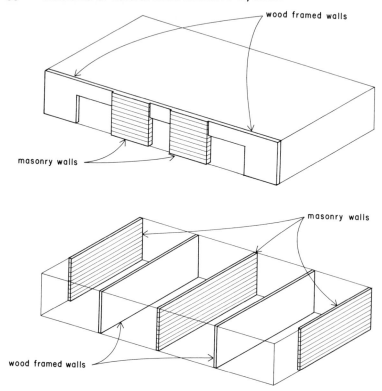

FIGURE 5.10 Mixed shear walls of varying stiffness.

do not suffer damage because of the wall distortions. An example of this situation is shown in Figure 5.11. The two relatively long solid portions are assumed to perform the bracing function for the entire wall and would be designed as isolated piers. However, when the wall deflects, the effect of the movement on the shorter piers, on the headers over openings, and on the door and window framing must be considered. The headers must not be cracked loose from the solid wall portions or pulled off their supports.

5.3 Braced Frames

Post and beam systems, consisting of separate vertical and horizontal members, may be inherently stable for gravity loading, but

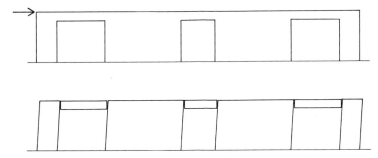

FIGURE 5.11   Effect of shear wall deflection on headers.

they must be braced in some manner for lateral loads. The three basic ways of achieving this are through shear panels, moment resistive joints between the members, or by trussing.

When shear panels are used, the panels themselves are usually limited to the direct shear force resistance. Elements of the frame are used for the purposes of diaphragm chords, collectors, drag struts, transfer members, and anchorage. Thus, the lateral resistive system is essentially that of a box system, although a complete frame structure exists together with the diaphragm elements of the box.

When moment resistive joints are used, the lateral loads induce bending and shear in the elements of the frame. The behavior of this type of system is discussed in Section 5.4. In some cases shear panels may also be added to the rigid frame, producing an interaction between the frame and the panels. The behavior of such interactive structures is discussed in Section 5.5.

The term *braced frame*, as used by the *UBC*, refers essentially to the use of trussing, or triangulation, of the frame to achieve its lateral stability. The trussing is usually formed by the insertion of diagonal members in the rectangular bays of the frame. If single diagonals are used, they must serve a dual function—acting in tension for the lateral loads in one direction, and in compression when the load direction is reversed. Because tension members are generally more efficient, the frame is sometimes braced with a double set of diagonals (called X-bracing) to eliminate the need for compression members. In any event, the trussing causes the lateral loads to induce only axial forces in the members of the frame, as compared to the behavior of the rigid frame. It also generally results in a frame that is stiffer for both static and dynamic loading, having less deformation than the rigid frame.

Some of the problems to be considered in using braced frames are the following:

1. Diagonal members must be placed so as not to interfere with the action of the gravity resistive structure or with other building functions. If the bracing members are designed essentially as axial stress members, they must be located and attached so as to avoid loadings other than those required for their bracing functions. They must also be located so as not to interfere with door, window, or roof openings or with ducts, wiring, piping, light fixtures, and so on.

2. As mentioned previously, the reversibility of the lateral loads must be considered. As shown in Figure 5.12, such consideration requires that diagonal members be dual-functioning (as single diagonals) or redundant (as X-bracing) with one set of diagonals working for load from

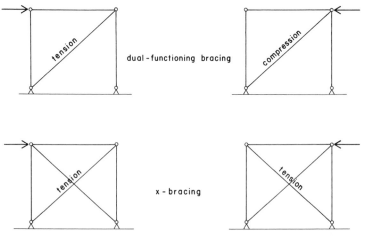

FIGURE 5.12   Dual-functioning trussing versus X-bracing.

one direction and the other set working for the reversal loading.

3. Although the diagonal bracing elements usually function only for lateral loading, the vertical and horizontal elements must be considered for the various possible combinations of gravity and lateral load. Thus, the total frame must be analyzed for all the possible loading conditions, and each member must be designed for the particular critical combinations that represent its peak response conditions.

4. Long, slender bracing members, especially in X-braced systems, may have considerable sag due to their own dead weight, which requires that they be supported by sag rods or other parts of the structure.

5. The trussed structure should be "tight." Connections should be made in a manner to assure that they will be initially free of slack and will not loosen under the load reversals or repeated loadings. This means generally avoiding connections that tend to loosen or progressively deform, such as those that use nails, loose pins, unfinished bolts, and so on.

6. In order to avoid gravity loading on the diagonals, the connections of the diagonals are sometimes made only after the gravity resistive structure is fully assembled and at least partly loaded by the building dead loads.

7. The deformation of the trussed structure must be considered, and it may relate to its function as a distributing element, as in the case of a horizontal structure, or to the establishing of its relative stiffness, as in the case of a series of vertical elements that share loads. It may also relate to some effects on nonstructural parts of the building, as was discussed for shear walls.

8. In most cases it is not necessary to brace every individual bay of the rectangular frame system. In fact, this is often not possible for architectural reasons. As shown in Figure 5.13, walls consisting of several bays can be braced by trussing only a few bays, or even a single bay, with the rest of the structure tagging along like cars in a train. Figure

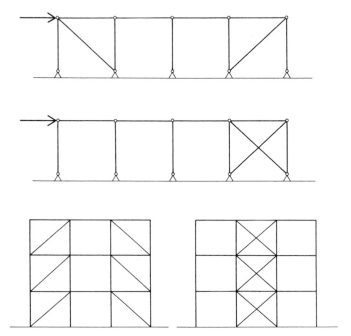

FIGURE 5.13 Partial trussing of framed structures.

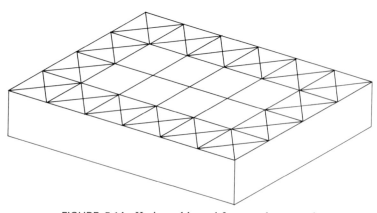

FIGURE 5.14 Horizontal braced frame—edge trussed.

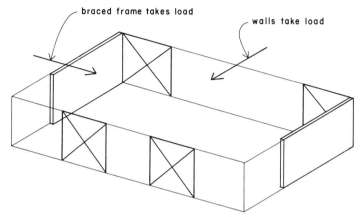

FIGURE 5.15   Mixed vertical elements for lateral resistance.

5.14 shows the case of a horizontally braced roof structure in which only the exterior bays are braced.

The braced frame can be mixed with other bracing systems in some cases. Figure 5.15 shows the use of a braced frame for the vertical resistive structure in one direction and a set of shear walls in the other direction. In this example the two systems act independently, except for the possibility of torsion, and there is no need for a deflection analysis to determine the load sharing.

Figure 5.16 shows a structure in which the end bays of the roof framing are X-braced. For loading in the direction shown, these braced bays take the highest shear in the horizontal structure, allowing the deck to be designed for a lower shear stress.

In general, trussing tends to produce a structure that has an overall stiffness somewhere between that of the stiff diaphragm and that of the flexible moment-resistive frame. Design criteria are generally based on the assumption that the braced frame is closer in stiffness to the diaphragm system. However, it is possible to have a fairly flexible diaphragm, such as a wood-framed one with a high depth-to-span ratio. It is also possible to produce a relatively rigid moment-resistive frame if the members are short and stiff.

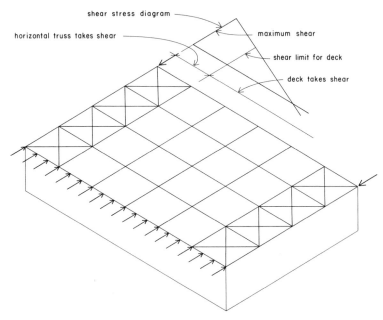

shear stress diagram

horizontal truss takes shear

maximum shear

shear limit for deck

deck takes shear

FIGURE 5.16   Mixed horizontal diaphragm and braced frame.

Section 2312(j)1G of the *UBC* gives some special requirements for the design of the members and connections for braced frames. These apply only for seismic loads on buildings in seismic zones 3 and 4.

5.4 Moment-resistive Frames

In rigid frames with moment-resistive connections, both gravity and lateral loads produce interactive moments between the members. The *UBC* requires that a rigid frame designed for seismic loading be classified as a "ductile moment-resistive space frame." Generally, frames of steel possess this character, but frames of reinforced concrete require special consideration of the reinforcing in order to meet this qualification.

In most cases rigid frames are actually the most flexible of the basic types of lateral resistive systems. This deformation character, together with the required ductility, makes the rigid frame a structure that absorbs energy loading through deformation, as well as through its sheer brute strength. The net effect is that the structure actually works less hard in force resistance, because its deformation tends to soften the loading. This is somewhat like rolling with a punch instead of bracing yourself to take it head on.

Most moment-resistive frames consist of either steel or concrete. Steel frames have either welded or bolted connections between the linear members to develop the necessary moment transfers. Frames of concrete achieve moment connections through the monolithic concrete and the continuity and anchorage of the steel reinforcing. Because concrete is basically brittle and not ductile, the ductile character is essentially produced by the ductility of the reinforcing. The type and amount of reinforcing and the details of its placing become critical to the proper behavior of rigid frames of reinforced concrete.

The design of moment-resistive ductile frames for seismic loads is beyond the scope of this book. Such design can be done only by using plastic design for steel and ultimate strength design for reinforced concrete. We present only a brief discussion of this type of structure. For wind loading the analysis and design may be somewhat more simplified. However, if the structure is considerably indeterminate, an accurate analysis may require a complex and laborious calculation. We show some examples in Chapters 6 and 7 but limit the analysis to approximate methods.

For lateral loads in general, the rigid frame offers the advantage of a high degree of freedom in architectural terms. Walls and interior spaces are freed of the necessity for solid diaphragms or diagonal members. For building planning as a whole, this is a principal asset. Walls, even where otherwise required to be solid, need not be of a construction qualifying them as shear walls.

When seismic load governs as the critical lateral load, the moment-resistive frame has the advantage of being required to carry less load than other types of lateral resistive systems. Table 23–I in the *UBC* gives the value of K for a rigid frame as 0.67 versus 1.33 for a box system with shear walls or braced frames. Thus, the rigid frame is required to carry only half the load specified for the box system. Even when the rigid frame is interactive with a box system, and the rigid frame is capable of carrying at least 25% of the lateral load, the K factor is reduced to 0.80, or 40% less than that for a box system alone.

Deformation analysis is a critical part of the design of rigid frames, because such frames tend to be relatively deformable when compared to other lateral resistive systems. The deformations have the potential of causing problems in terms of movements of a disturbing nature that can be sensed by the building occupants or of damage to nonstructural parts of the building, as previously discussed. The need to limit deformations often results in the size of vertical elements of the frame being determined by stiffness requirements, rather than by stress limits.

A special type of rigid frame is one produced by a plywood and timber construction in which plywood is nailed to two sides of a wood frame to produce box sections with plywood webs and timber flanges and stiffeners. This system may be used to achieve a spanning rigid frame, or it may simply be a means of transforming a wall structure into a continuous rigid frame, rather than a series of connected shear walls, for lateral load resistance.

5.5 Interactive Frames and Diaphragms

Most buildings have some solid walls, that is, walls with continuous surfaces free of openings. When the gravity load resistive structure of the building consists of a frame, the relationship between the walls and the frame has several possibilities with regard to action caused by lateral loads.

The frame may be a braced frame or a moment-resistive frame designed for the total resistance of the lateral loads, in which case the attachment of walls to the frame must be done in a manner that prevents the walls from absorbing lateral loads. Because solid walls tend to be quite stiff in their own planes, such attachment often requires the use of separation joints or flexible connections that will allow the frame to deform as necessary under the lateral loads.

The frame may be essentially designed for gravity resistance only, with lateral load resistance provided by the walls acting as shear walls. This method requires that some of the elements of the frame function as collectors, stiffeners, shear wall end members, or diaphragm chords. If the walls are intended to be used strictly for lateral bracing, care must be exercised in the design of construction details to assure that beams that occur above the walls are allowed to deflect without transferring loads to the walls.

When walls are firmly attached to vertical elements of the frame, they usually provide continuous lateral bracing in the plane of the wall, thus permitting the vertical frame elements to be designed for column action using their stiffness in the direction perpendicular to the wall. Thus 2 × 4 studs may be designed as columns using h/d ratios based on their larger dimension.

In some cases both walls and frames may be used for lateral load resistance at different locations or in different directions. Figure 5.17 shows two such situations. In the upper part of the figure a shear wall is used at one end of the building and a frame at the other end for the wind from one direction. In the lower part of the figure walls are used for the lateral loads from one direction and frames for the load from the other direction. In both cases the walls and frames do not actually interact; that is, they act independently with regard to load sharing.

Figure 5.18 shows two situations in which walls and frames interact to share a direct load. In the upper part of the figure the interior walls and the end frames share the total load from a single direction. If the horizontal structure is a rigid diaphragm, the load sharing will be on the basis of the relative stiffness of the vertical elements. This relative stiffness must be established by the calculated deflection resistance of the elements, as previously discussed.

The lower part of Figure 5.18 shows a situation in which the walls and frames interact to share a single direct load. This represents a highly indeterminate situation. Case 3 in Table 23–I of the *UBC* describes such a situation, called a "dual bracing system," that consists of shear walls and a moment-resistive frame. The *UBC* requires that the shear walls in this system be capable of resisting the entire lateral load but that the frame be capable of resisting 25% of the load by itself. This is an apparent

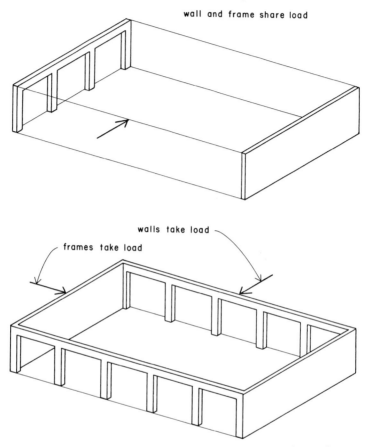

wall and frame share load

walls take load

frames take load

FIGURE 5.17 Mixed shear walls and rigid frames—no interaction.

redundancy in the design, but the advantage gained is that the K factor for the seismic load may be reduced from 1.33 for the walls alone to 0.80 for the dual system.

Structures of the type illustrated in Figure 5.18 should be analyzed using dynamic methods to determine the load distribution and ultimate or plastic strength analysis for the behavior of the elements of the system. If the simpler equivalent static load method is used and the elements are designed using working

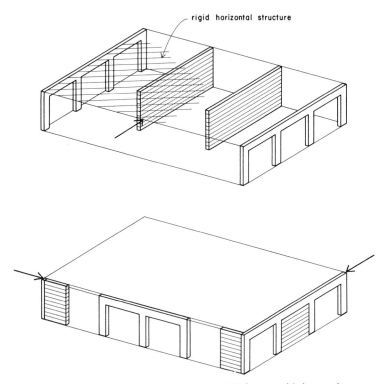

rigid horizontal structure

FIGURE 5.18 Mixed shear walls and rigid frames with interaction.

stress analysis, the stiffer elements of the system should be designed quite conservatively, because they will tend to take more loading than the static analysis implies.

5.6 Collectors

Transfer of loads from horizontal to vertical elements in laterally resistive structural systems frequently involves the use of some structural members that serve the functions of struts, drags, ties, collectors, and so on. These members often serve two functions—as parts of the gravity resistive system or for other functions in lateral load resistance.

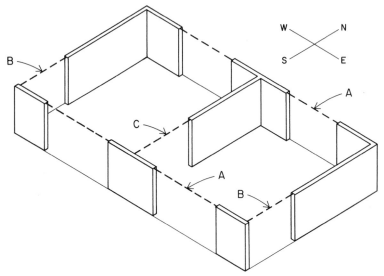

FIGURE 5.19   Use of framing members in the box building.

Figure 5.19 shows a structure consisting of a horizontal dia-phragm and a number of exterior shear walls. For loading in the north–south direction the framing members labeled "A" serve as chords for the roof diaphragm. In most cases they are also parts of the roof edge or top of the wall framing. For the lateral load in the east–west direction they serve as collectors. This latter func-tion permits us to consider the shear stress in the roof diaphragm to be a constant along the entire length of the edge. The collector "collects" this constant stress from the roof and distributes it to the isolated shear walls, thus functioning as a tension/compres-sion member in the gaps between the walls.

In the example in Figure 5.19 the collector "A" must be at-tached to the roof edge to develop the transfer of the constant shear stress. The collector "A" must be attached to the indi-vidual shear walls for the transfer of the total load in each wall. In the gaps between walls the collector gathers the roof edge load and functions partly as a compression member, pushing some of the load to the forward wall, and partly as a tension member, dragging the remainder of the collected load into the rearward wall.

Collectors "B" and "C" in Figure 5.19 gather the edge load from the roof deck under the north–south lateral loading. Their function over the gap reverses as the load switches direction. They work in compression for load in the northerly direction, pushing the load into the walls. When the load changes to the southerly direction, they work in tension, dragging the load into the walls.

The complete functioning of a lateral resistive structural system must be carefully studied to determine the need for such members. As mentioned previously, ordinary members of the building construction will often serve these functions: top plates of the stud walls, edge framing of roofs and floors, headers over openings, and so on. If so used, such members should be investigated for the combined stress effects involved in their multiple roles.

5.7   Anchorage Elements

The attachment of elements of the lateral resistive structure to one another, to collectors, or to supports usually involves some type of anchorage element. There is a great variety of these, encompassing the range of situations with regard to load transfer conditions, magnitude of the forces, and various materials and details of the structural members and systems. Many of these elements and their installation details are illustrated in the examples in Chapters 6 and 7. The following is a discussion of some of the basic types of elements and anchorage situations.

Tie Downs.   Resistance to vertical uplift is sometimes required for elements of a braced or moment resistive frame, for the ends of shear walls, or for light roof systems subject to the force of upward wind suction. For concrete and reinforced masonry structures, such resistance is most often achieved by doweling and/or hooking of reinforcing bars. Steel columns are usually anchored by the anchor bolts at their bases. The illustrations in Figure 5.20 show some of the devices that are used for anchoring wood structural elements. In many cases these devices have been

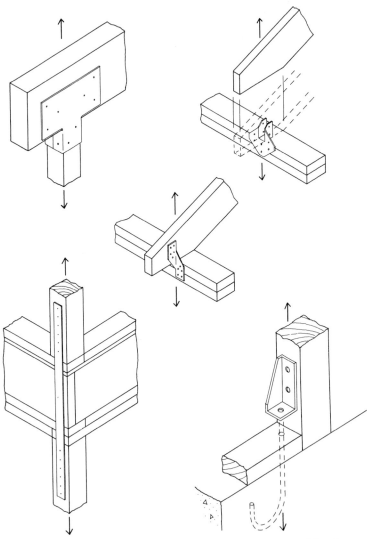

FIGURE 5.20 Anchorage devices for uplift resistance—wood framing.

load tested and their capacities rated by their manufacturers. When using them, it is essential to determine whether the load ratings have been accepted by the building code agency with jurisdiction for a specific building design.

Horizontal Anchors. In addition to the transfer of vertical gravity load and lateral shear load at the edges of horizontal diaphragms, there is usually a need for resistance to the horizontal pulling away of walls from the diaphragm edge. In many cases the connections that are provided for other functions also serve to resist this action. Codes usually require that this type of anchorage be a "positive" one, not relying on such things as the withdrawal of nails or lateral force on toe nails. Figure 5.21 shows some of the means used for acheiving this type of anchorage.

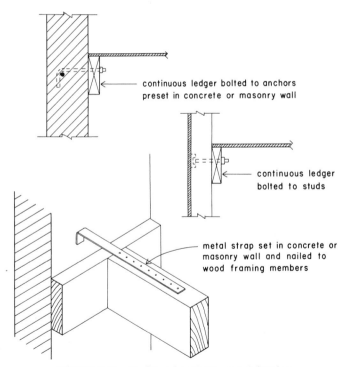

continuous ledger bolted to anchors preset in concrete or masonry wall

continuous ledger bolted to studs

metal strap set in concrete or masonry wall and nailed to wood framing members

FIGURE 5.21   Horizontal anchors—wood framing.

Some additional details for various situations are shown in the examples in Chapters 6 and 7. Some of the patented devices used for this, and other types of anchorage, are shown in the Appendix.

Shear Anchors. The shear force at the edge of a horizontal diaphragm must be transferred from the diaphragm into a collector or some other intermediate member, or directly into a vertical diaphragm. Except for poured-in-place concrete structures, this process usually involves some means of attachment. For wood structures the transfer is usually achieved through the lateral loading of nails, bolts, or lag screws, for which the codes or industry specifications provide tabulated load capacities. For steel deck diaphragms the transfer is usually achieved by welding the deck to supporting steel framing. If the vertical system is a steel frame, these members are usually parts of the frame system. If the vertical structure is concrete or masonry, the edge transfer members are usually attached to the walls with anchor bolts set in the concrete or in solid-filled horizontal courses of the masonry. As in other situations, the combined stresses on these connections must be carefully investigated to determine the critical load conditions.

The complete transfer of force from the horizontal to the vertical elements of the lateral resistive system can be quite complex in some cases. Figure 5.22 shows a joint between a horizontal plywood diaphragm and a vertical plywood shear wall. For reasons other than lateral load resistance, it is desired that the studs in the wall run continuously past the level of the roof deck. This necessitates the use of a continuous edge-framing member, called a ledger, that serves as the vertical support for the deck as well as the chord and edge collector for the lateral forces. This ledger is shown to be attached to the faces of the studs with two lag screws at each stud. The functioning of this joint involves the following:

1. The vertical gravity load is transferred from the ledger to the studs directly through lateral load on the lag screws.
2. The lateral shear stress in the roof deck is transferred to the ledger through lateral load on the edge nails of the deck. This stress is in turn transferred from the ledger to the

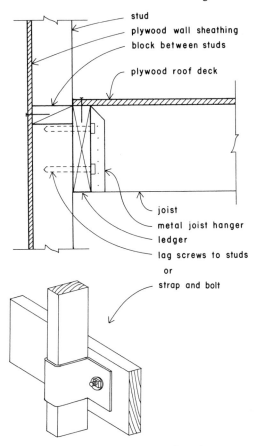

stud

plywood wall sheathing

block between studs

plywood roof deck

joist

metal joist hanger

ledger

lag screws to studs

or

strap and bolt

FIGURE 5.22   Joint between horizontal and vertical elements—wood box system.

studs by horizontal lateral load on the lag screws. The horizontal blocking is fit between the studs to provide for the transfer of the load to the wall plywood, which is nailed to the blocking.

3. Outward loading on the wall is resisted by the lag screws in withdrawal. This is generally not considered to be a good positive connection, although the load magnitude should be

considered in making this evaluation. A more positive connection is achieved by using the bolts and straps shown in the lower sketch in Figure 5.22.

Another shear transfer problem is that which occurs at the base of a shear wall in terms of a sliding effect. For a wood-framed wall some attachment of the wall sill member to its support must be made. If the support is wood, the attachment is usually achieved by using nails or lag screws. If the support is concrete or masonry, the sill is usually attached to preset anchor bolts. The lateral load capacity of the bolts is determined by their shear capacity in the concrete or the single shear limit in the wood sill. *UBC* Section 2907(e) gives some minimum requirements for sill bolting, which should be used as a starting point in developing this type of connection.

For walls of concrete and masonry, in which there is often considerable dead load at the base of the wall, sliding resistance may be adequately developed by friction. Doweling provided for the vertical wall reinforcing also offers some lateral shear resistance. If a more positive anchor is desired, or if the calculated load requires it, shear keys may be provided by inserting wood blocks in the concrete, as shown in Figure 5.23.

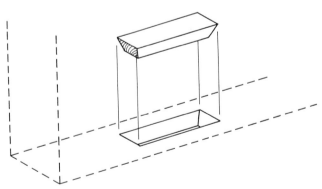

FIGURE 5.23   Forming a shear key for a concrete or masonry wall.

5.8 Foundations

Building foundations are important parts of the lateral load resistive system. For wind load the final resolution of the wind force is into the ground, which usually involves some combination of overturn, horizontal sliding, and the development of passive lateral soil pressure. For seismic load the foundations have a dual role. Initially, they are the origin point for load on the building, being directly attached to the ground and used by the ground to shake the building. In our analysis procedures, however, the seismic load is considered to be a result of the inertial effect of the moving building mass. From this viewpoint, the loading condition becomes similar to that for the wind load. Thus, we visualize the horizontal inertial force as being transmitted through the building structure, into the foundation, and finally into the ground, which is now considered to offer passive resistance, as it does actually to the wind force.

Foundation design problems have considerable variety because of the wide range of possible soil conditions, load magnitudes, building size and shape, and type of structural system. For any building it is wise to have a subsoil investigation, lab tests on representative soil samples, and a recommendation from an engineer with experience in soil behaviors and foundation problems. We illustrate some of the ordinary and simple design problems in the examples in Chapters 6 and 7 but do not attempt to deal with all the special problems that can occur in the design of foundations for lateral loads.

For seismic loads, it is usually desirable that the entire building foundation act essentially as a single rigid unit. If elements of the foundation are isolated from one another, as in the case of individual column footings, it may be necessary to provide struts or grade walls to tie the structure into a unit. Where they exist, of course, the ordinary elements such as basement walls, grade walls, wall footings, and grade-level framing members may be used for this tying function.

For buildings with exceptional uplift or overturn effects, the foundation must usually provide an anchor through its own sheer

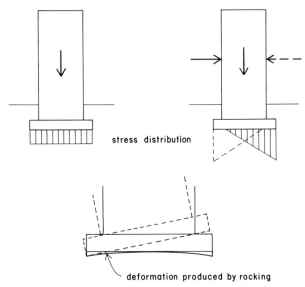

stress distribution

deformation produced by rocking

FIGURE 5.24 Soil deformation resulting from a lateral load.

dead weight. For a large building with shallow foundations and no basement, the necessary dead load may not exist in the foundations designed for gravity loads alone. This situation may require some additional mass in the foundation itself or the use of soil weight in the form of backfill over the foundation.

Another problem with shallow foundations is that they often bear on relatively compressible soil. Figure 5.24 illustrates the effect that can occur with a large lateral load and a strong overturning moment. With repeated applications and reversals of such a load, the soil beneath the foundation edges become compressed, resulting in an increasing tendency for the structure to rock, thus producing a loss of stability or a significant change in the dynamic load behavior of the structure. When the bearing strata of soil are very compressible, it is generally advisable to avoid the extreme condition of soil stress shown in the illustration.

Chapter 29 of the *UBC* provides some data and recommendations for use in designing for sliding or lateral passive

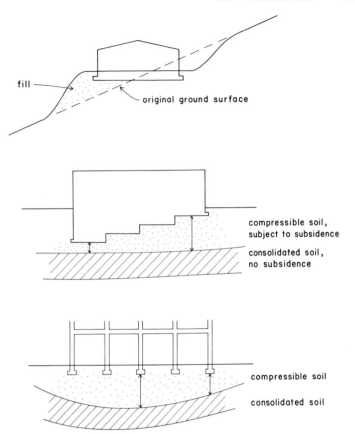

fill

original ground surface

compressible soil, subject to subsidence

consolidated soil, no subsidence

compressible soil

consolidated soil

FIGURE 5.25 Potential foundation problems.

resistance of soil. The *UBC* material is quite conservative in most cases, and a complete soils investigation and report may provide more rational data and permit a less conservative design. In many cases, soil conditions are general local phenomena, and local building codes often have special requirements and procedures for foundation design. Local design and construction procedures are sometimes based more on histories of successes and failures than on scientific analysis or engineering judgment.

Buildings on deep foundations consisting of long piles or piers offer special problems in design for lateral loads. The entire lateral force must usually be developed by the passive resistance of soil, because the sliding friction present on the bottom of shallow bearing footings is not available. When the upper soil strata are considerably deformable, the fundamental period of the building and its general dynamic behavior may be significantly affected by the lateral movement of the foundation. It is highly recommended that the design of these foundations and the structures supported by them be done by an engineer with experience in these problems.

Many seismic load failures of building structures are precipitated or aggravated by soil movements in fill or other highly compressible soil deposits. An especially hazardous situation is that which occurs when different parts of the foundation are placed on soil of significantly different compressibility, as may occur because of a hilly site, of variations in the level of the foundations, of extensive regrading, or of nonhorizontal soil strata. Subsidence of the fill, major differences in settlements, or lateral movement of the soil mass can produce various critical situations. Figure 5.25 illustrates some of the possibilities that can produce structural failures for both gravity and lateral load conditions.

6

Analysis and Design for Wind

||||||||||||||||||||||||||||||||||||||||||||||||||||||

This chapter consists of examples of analysis and design for wind in the design of specific types of buildings and other structures. The examples cover a wide range of situations with regard to building size, type of construction, and the problems of wind effect. Reference should be made to Chapters 1 through 4 for explanation of basic concepts, typical design procedures, and code requirements. A general discussion of basic elements of lateral resistive systems is given in Chapter 5, and some special problems are discussed in Chapter 8.

There is, of course, a need for coordination of the analysis and design for wind and that for gravity loads. For the sake of brevity, we have presented only a minimal discussion of the design for gravity, generally using approximate results without showing complete calculations.

Many of the examples consist of minor variations of preceding examples in order to permit comparisons and to shorten the work. When a variation of a previous example is used, the work shown is usually limited to the material pertinent to the changes, and reference should be made to the original example for a more complete discussion of the design.

Some of the example cases in this chapter are also used as examples for seismic analysis and design in Chapter 7. This is done primarily for brevity, but it also permits some comparisons between the work and the results for the two load conditions.

6.1   Example 1

We begin with a simple example: a one-story box with a horizontal roof diaphragm and vertical, exterior bracing walls. A plan and a section of the building are shown in Figure 6.1. The construction is light wood frame with a plywood roof deck and planar surfacing materials on wood studs for the walls.

We assume that the *UBC* wind map (Figure 4, Chapter 23 of the *UBC*; see also the Appendix) requires a basic wind pressure of 25 lb/ft². From Table 23–F of the *UBC* we obtain a design pressure of 20 lb/ft² for the building, whose total height is under 30 ft.

Assuming the roof structure to be at approximately 12 ft above the floor at the front of the building, the wind force applied to the edge of the roof diaphragm with the wind blowing north will be as shown in Figure 6.2. The assumption is made that the wall spans vertically and that the roof edge carries half of the span plus the 2 ft of cantilevered parapet. With these assumptions, the load is

$$\text{Total } H = (20 \text{ lb/ft}^2)(8 \times 50) = 8000 \text{ lb}$$

Or, as a uniform linearly distributed load on the roof edge, the load is

$$w = (20)(8) = 160 \text{ lb/ft}$$

The action of the roof deck is shown in Figure 6.3. The roof spans from end to end of the building, transferring its end reactions to the two 18-ft-long end shear walls. The maximum shear stress in the roof deck at the building ends is

$$v = \frac{4000}{30} = 133 \text{ lb/ft}$$

The maximum force in the edge chord members due to the moment at the center of the diaphragm span is

$$T = C = \frac{50,000}{30} = 1667 \text{ lb}$$

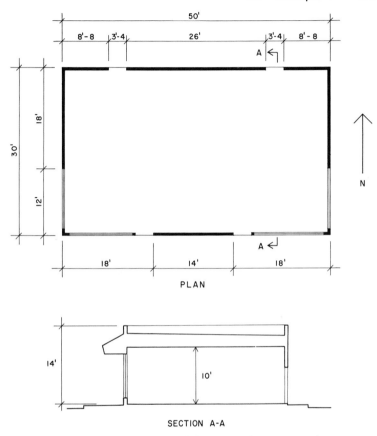

PLAN

SECTION A-A

FIGURE 6.1   Building (Example 1).

The load on the end shear walls is as shown in Figure 6.4. Because of the large opening, the design length of the shear wall is assumed to be 18 ft, although the structure over the opening is assumed to function in collecting load from the end of the roof. The combined wind and gravity loads on the shear wall are shown in Figure 6.5. The gravity load consists of the weight of the wall and some portion of the dead load of the roof and ceiling construction, assuming that the framing results in some distribution of these loads to the wall. We assume the total dead load to be 6000 lb.

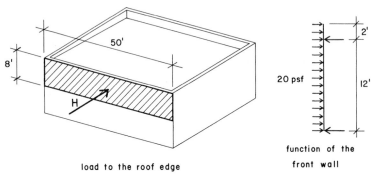

50'

8'

H

2'

20 psf

12'

function of the
front wall

load to the roof edge

FIGURE 6.2 Front wall function and roof edge load for north–south wind (Example 1).

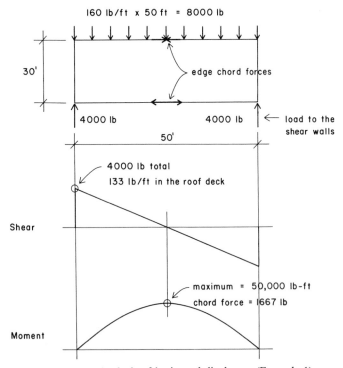

160 lb/ft x 50 ft = 8000 lb

30'

edge chord forces

4000 lb

4000 lb

← load to the
shear walls

50'

4000 lb total

133 lb/ft in the roof deck

Shear

maximum = 50,000 lb-ft

chord force = 1667 lb

Moment

FIGURE 6.3 Analysis of horizontal diaphragm (Example 1).

112

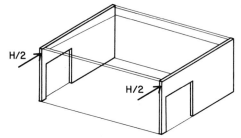

FIGURE 6.4   Loads to the end shear walls (Example 1).

The maximum shear stress in the wall is

$$v = \frac{4000}{18} = 222 \text{ lb/ft}$$

Using the *UBC* requirement for a safety factor of 1.5 against overturn, the moment to be resisted is

Overturning M: (4000)(12)(1.5) = 72,000 lb-ft

The gravity load resistance is

Gravity M: (6000)(9) = 54,000 lb-ft

which leaves a net moment of 18,000 lb-ft to be resisted by the end anchorage of the wall. The force on this end anchorage is calculated as

$$T = \frac{18,000}{18} = 1000 \text{ lb}$$

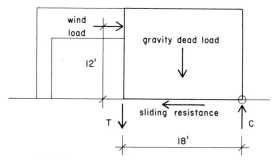

FIGURE 6.5   Shear wall actions (Example 1).

This force may be supplied by an end anchorage device (called a *hold down* or *tie down*), which is bolted to the end framing of the wall and to an anchor bolt preset in the concrete base. Providing such a device in this case may be quite conservative, however, because there are other elements providing hold-down effects on the ends of the shear wall. If the opening is spanned by a header, the end reaction of the header will deliver an additional dead load to the end framing of the wall, which may exceed the required 1000-lb force. At the corner of the building, the plywood on the two intersecting walls will probably be nailed to some common end framing. Thus, the end of the rear wall provides some additional dead load resistance at this point.

The end framing members in the 18-ft shear wall should be designed for the combination of the gravity loads plus the compression that is the result of the overturning moment. The overturning moment for this calculation is

$$\text{Overturning } M: (4000)(12) = 48,000 \text{ lb-ft}$$

and the resulting compression force from wind alone is

$$C = \frac{48,000}{18} = 2667 \text{ lb}$$

Horizontal sliding of the shear wall must be resisted by the bolts that secure the sill plate to the foundation. *UBC* Section 2907(e) requires that these bolts be a minimum of 1/2 in. in diameter, a maximum of 6 ft on center, and that there be a bolt not more than 12 in. from each end of the wall. For the 18-ft-long wall these requirements would result in a minimum of four 1/2 in. bolts. The lateral load capacity of each bolt will be limited by the single shear in the wood sill or the rated resistance of the concrete. For a 2× sill of Douglas fir, *UBC* Table 25–F gives an allowable load value of 650 lb for a 1/2-in. bolt in single shear, parallel to the grain. This requirement is compared with the value from *UBC* Table 26–G, which gives a limit of 2000 lb for a 1/2-in. bolt in shear with an ultimate concrete strength of either 2000 or 3000 psi. Using the lower stress limit of the wood and adjusting for the allowable stress increase with wind load, the total capacity of the four sill bolts is

$$V = (1.33)(4)(650) = 3458 \text{ lb}$$

which is slightly less than the required 4000 lb and requires the addition of another 1/2-in. bolt or an increase in the bolt size.

The diaphragm shear stresses in the roof and end walls are well within the capacity of ordinary structural plywood. Assuming 1/2-in. plywood for the roof deck and 3/8-in. plywood for the wall surfacing, blocked diaphragms with the minimum nail spacing of 6 in. at panel edges and diaphragm boundaries will safely carry the loads. (See *UBC* Tables 25–J and 25–K in the Appendix.)

It is also possible to consider other surfacing for the end shear walls. If cement plaster is used on both surfaces (as plaster on the inside and stucco on the outside), its single surface value of 180 lb/ft taken from *UBC* Table 47–I can be doubled, giving a total capacity of 360 lb/ft, which is well above the required shear of 222 lb/ft as previously calculated for the wall. It should be noted that *UBC* Section 4713(a) permits this addition of the two surfaces only when they are covered with the same materials.

The construction details at the intersection of the roof and wall must be carefully developed to facilitate the various load transfers at this point. There are various possibilities for these details, which must be developed to satisfy the lateral load and gravity load transfers and the desired architectural features. Figure 6.6 shows one possibility for this connection. Because the roof level varies along the wall length, a ledger is bolted to the face of the studs to provide support for the roof deck. This procedure permits the studs to be a constant length and the top plates of the wall to be horizontal. The nailing of the edge of the roof deck to the ledger, the attachment of the ledger to the studs, and the nailing of the wall plywood to the studs and blocking must all be designed for the calculated stress of 133 lb/ft at the roof edge.

If the roof joists are parallel to this wall, the vertical gravity load on the ledger will be small. If the joists frame into the ledger, as shown in Figure 6.6, the load will be somewhat higher. In either case the lateral load on the lag screws must be designed for the combination of the vertical gravity and the horizontal wind effects. In addition, they must resist the outward suction force on the wall by withdrawal. There are various other devices that may be used to attach the ledger to the studs for a more positive resistance to the outward force. (See Figure 5.22.)

One type of hold-down device for the ends of the shear wall is

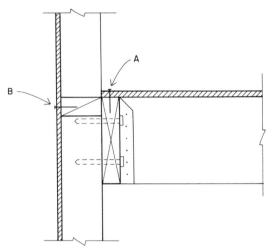

FIGURE 6.6 Roof/wall joint (Example 1, end wall).

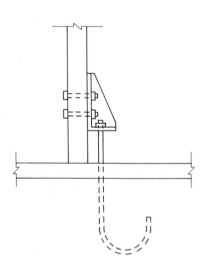

FIGURE 6.7 Typical hold-down device for a wood shear wall.

116

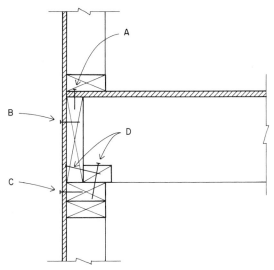

FIGURE 6.8 Roof/wall joint (Example 1, front wall).

shown in Figure 6.7. The shear on the bolts to the end framing member, the tension and embedment of the anchor bolt, and the stresses in the device must all resist the hold-down force. Common practice is to use a manufactured device that has been load tested and rated with code-approved loadings given by the manufacturer. (See the Appendix for some patented devices.)

Figure 6.8 shows a possible detail for the connection of the roof to the wall at the front of the building. This connection is also subject to considerable variation, depending on the desired details of the parapet, the canopy structure, and the window head at this location. In the example shown, the roof framing sits directly on the top plates of the stud wall, and the parapet is built as a stub wall on top of the roof deck. Nails "A" and "B" effect the transfer of the east–west wind load from the roof to the front shear walls.

The outward wind load on the wall may be carried back to the roof structure in a number of ways. The method illustrated in Figure 6.8 is that of using the continuous ribbon to drag the load into the top plates of the wall. The roof deck is nailed to this member, and the plates are secured to it by the horizontal blocks and nails "D." An alternative would be to use metal framing anchors to secure the joists to the plates.

There are three elements in this detail shown in Figure 6.8 that may contribute to the development of the chord for the roof diaphragm: the top plates of the lower wall, the continuous ribbon, or the sill of the parapet. The relatively low chord force in this example could be developed by any one of these alone; thus, their combined effect is quite redundant if all of them work. The critical consideration is the continuity of the chord force through the splices in the members, because each of them would be composed of several separate pieces in the 50-ft-long building. Assuming it to be unlikely that all of these members would be spliced at the same location, their overlapping should effectively provide the necessary continuity for the chord. If a positive splice is deemed necessary, one of the members—most likely the ribbon or the top plates—could be spliced with a bolted joint.

East–west wind on the building produces a load on the edge of the roof as shown in Figure 6.9. The vertical dimension of the load zone depends on the height of the roof and the nature of the spanning action of the walls, but for simplification in the example it is assumed to be the same as for the front wall. The direct action of the roof in spanning between the front and rear walls is shown in Figure 6.10. It is assumed that the diaphragm acts as a simple beam with equal reactions applied to the front and rear walls. On this basis the total wind load is

$$H = (20 \text{ lb/ft}^2)(8)(30) = 4800 \text{ lb}$$

and the stresses are

Maximum shear in the roof:

$$v = \frac{2400}{50} = 48 \text{ lb/ft}$$

Chord force:

$$T = C = \frac{18,000}{50} = 360 \text{ lb}$$

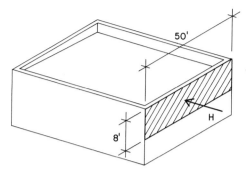

FIGURE 6.9   Roof edge load for east–west wind (Example 1).

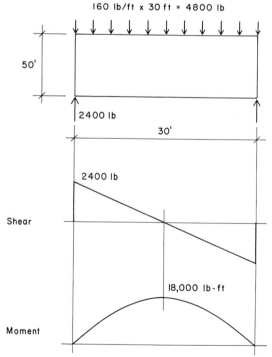

FIGURE 6.10   Horizontal diaphragm functions for east–west wind (Example 1).

119

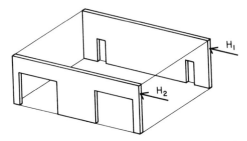

FIGURE 6.11   Loads to the front and rear shear walls (Example 1).

Shear in the front wall:

$$v = \frac{2400}{14} = 171 \ \text{lb/ft}$$

Average shear in the rear wall:

$$v = \frac{2400}{26 + 2(8.67)} = 55 \ \text{lb/ft}$$

As shown in Figure 6.11, the total horizontal force will be distributed between the front and rear walls. In the preceding analysis, it is assumed that the roof functions in simple spanning and the two forces, H_1 and H_2, are equal. The two walls are not symmetrical, however, and their difference in stiffness will tend to cause a torsion effect that will increase the force on the front wall and slightly reduce that on the real wall. The torsion will be resisted by all the shear walls acting to resist the rotation effect on the building.

Figure 6.12 illustrates the basis for calculation of the torsional effect. Individual shear walls are assumed to have a stiffness in proportion to their plan length (the usual assumption for wood-framed walls.) The center of stiffness is located by a simple static moment calculation as follows:

$$\bar{y} = \frac{\text{moment of the front wall about the rear wall}}{\text{the sum of stiffnesses of the two walls}}$$

$$= \frac{14 \times 30}{57.33} = 7.33 \ \text{ft}$$

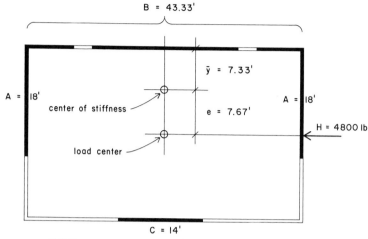

FIGURE 6.12   Torsion due to east–west wind (Example 1).

The torsional moment of inertia of the shear walls, J, is found as shown in Table 6.1. The torsional moment is determined as:

$$T = (4800)(7.67) = 36816 \text{ lb-ft}$$

The critical stress on the front wall is now determined by adding the torsional stress to the direct load stress. The torsional stress is calculated as

$$v = \frac{Tc}{J} = \frac{(36816)(22.67)}{32023} = 26 \text{ lb/ft}$$

TABLE 6.1   Torsional Resistance of the Plywood Shear Walls

Wall	Wall Stiffness (S) (plan length in feet)	Distance from Center of Stiffness (in feet)	$J = Rd^2$
A	2(18)	25	22,500
B	43.33	7.33	2,328
C	14	22.67	7,195
	Total torsional moment of inertia (J) =		32,023 ft^4

The direct stress for this calculation is found by dividing the total H force by the sum of stiffnesses of the front and rear walls. Thus

$$v = \frac{4800}{57.33} = 84 \text{ lb/ft}$$

and the total stress in the front wall is

$$v = 26 + 84 = 110 \text{ lb/ft}$$

This result is lower than the value of 171 lb/ft as determined by the previous analysis. For the front wall, therefore, the conservative assumption would be to ignore the torsional effect and use the peripheral analysis in which H_1 is assumed to equal H_2.

For the rear wall the torsional effect works to reduce the shear. However, most codes do not allow this reduction, so the critical value for the rear wall is either the 55 lb/ft from the peripheral analysis or the 84 lb/ft from the direct force analysis using the total wall stiffnesses.

A conservative procedure would be to perform both analyses and to use the highest values from either analysis for the individual elements of the system. If plywood walls are used in the example, this judgement is academic, because all the calculated values are well below the minimum capacity of ordinary plywood shear walls.

With an assumed total dead load of 6000 lb on the front wall, the higher value of 171 lb/ft for shear produces the situation for overturn and sliding as shown in Figure 6.13. The overturn is analyzed as follows:

Lateral load:	$171(14) = 2400$ lb (peripheral basis)		
Overturn M:	$2400(12 \text{ ft})(1.5 \text{ } SF)$	=	43,200 lb-ft
Dead load M:	$6000(7 \text{ ft})$	=	42,000
Net moment for hold down:		=	1,200 lb-ft

This is a small force. Header loads on the ends of the wall will provide additional dead load so that anchored hold downs are probably not required for this wall.

The sill for the wall will be bolted with a minimum of three

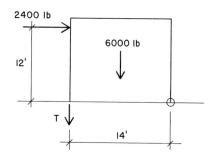

2400 lb

6000 lb

12'

T

14'

FIGURE 6.13 Overturn analysis for front shear wall (Example 1).

bolts, which will be adequate for the sliding resistance, even if the minimum 1/2-in. bolts are used. (See previous discussion for the end shear wall.)

There are two approaches to the design of the rear wall. The first approach is to assume that the wall consists of three isolated piers with heights equal to the floor-to-roof distance, as is shown as Case 1 in Figure 6.14. The overturn analysis (see Figure 6.15) of the short pier on this basis is as follows:

Lateral load:	$84(8.67) = 728$ lb (direct load basis)		
Overturn M:	$728(11$ ft$)(1.5$ $SF)$	$=$	12,012 lb-ft
Dead load M:	$3000(4.33$ ft$)$	$=$	12,990
Net moment for hold down:			0

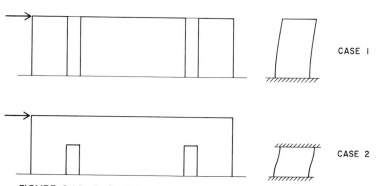

CASE 1

CASE 2

FIGURE 6.14 Optional functioning of the rear shear wall (Example 1).

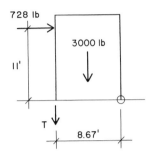

FIGURE 6.15 Overturn analysis for the rear shear wall (Example 1).

For case 2 in Figure 6.14, the continuous portion of the wall above the openings is assumed to produce a rigid-frame action, which results in the short piers being fixed at both their tops and bottoms, instead of cantilevering as in Case 1. This results in a double reduction of the moment at the base of the piers. The first reduction is due to the reduced pier height: from 11 ft to 7 ft. The second reduction is due to the fixity at both top and bottom. If the inflection of the pier is assumed at midheight, as shown in Figure 6.16, the moment arm for the overturn is reduced to 3.5 ft, which will considerably reduce the overturn effect at the base of the piers, while the resisting moment of the dead load remains the same.

There are some features required of the construction details for both of these cases. For Case 1 the end framing members of the piers must be continuous from floor to roof for continuity of the

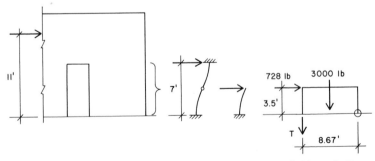

FIGURE 6.16 Optional assumption for the rear shear wall (Example 1).

end forces, as shown in Figure 6.17. Because this practise is ordinarily followed in the construction of stud walls, it does not constitute any addition for the lateral load effect. For Case 2 there must be both vertical and horizontal continuity of the framing at the top corners of the door openings. This is necessary for the development of the rigid frame. The vertical members would thus be the same as for Case 1, but the horizontal continuity must be achieved by the addition of horizontal blocking between the studs and a tension tie across the corner, as shown in the figure.

Assuming the inflection point of the short pier in Case 2 to be at its midheight, as shown in Figure 6.16, the moment due to lateral load at both the top and bottom of the pier is

$$M = (728)(3.5) = 2548 \text{ lb-ft}$$

which is used for the design of the tension tie at the corner of the opening and for the overturning moment at the base.

The overturn analysis for Case 2, as shown in Figure 6.16, thus becomes

Overturn M:	$(2548)(1.5 \, SF)$	=	3822 lb-ft
Dead load M:	(As for Case 1)	=	12990
Net moment for hold down:			0

Because both cases in this example result in no requirement for a hold down at the base, it would seem to be wise to assume Case 1 and omit the blocks and straps. However, many designers prefer to use this detail at openings routinely in order to reduce the possibility of vertical cracking of the wall at the corner of the opening.

For all of the walls, the framing at the roof edge or at the top of the wall must function as a collector, as discussed in Section 5.6. At the end wall, the collector over the 12-ft-long opening accumulates a total load from the roof edge of

$$F = (133 \text{ lb/ft})(12 \text{ ft}) = 1596 \text{ lb}$$

For wind in a northerly direction, this force is pushed into the shear wall, and the collector functions as a compression member. For load in a southerly direction, the load is dragged into the shear wall, and the collector functions as a tension member. The

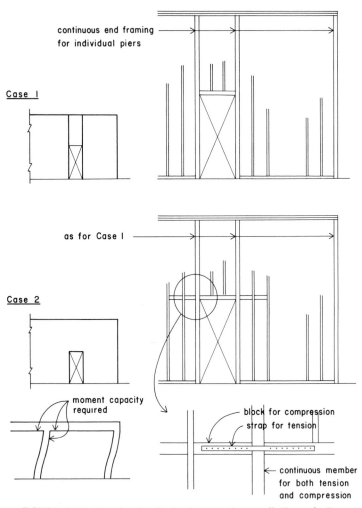

continuous end framing
for individual piers

Case 1

as for Case 1

Case 2

moment capacity
required

block for compression
strap for tension

continuous member
for both tension
and compression

FIGURE 6.17   Framing details for the rear shear wall (Example 1).

126

framing that is considered to perform these functions would be designed for both of these structural actions. Most likely the compression would be critical for the framing member, and the tension would be considered only if the member must be spliced.

If the end wall is constructed as shown in Figure 6.6, the continuous ledger would be considered to perform these functions. This member is more than adequate for the compression force, and if a splice can be avoided over the wall opening, no special consideration would be necessary in the construction for the collector function.

In addition to the main elements of the lateral resistive system, there are various other problems to be dealt with in design for wind effects on the building. These include the following:

Design of the wall studs for combined gravity and wind.

Design of the parapet.

Design of the canopy for uplift.

Design of the roof/wall connection for uplift.

Design of window and door framing and glazing for wind.

Design of the attachments for roof or wall mounted signs, equipment, and so on.

With this type of construction, the wall and parapet are usually either as shown in Figure 6.6 or as shown in Figure 6.8. If the studs are continuous above the roof, as shown in Figure 6.6, they become analogous to a beam with one overhanging end and their behavior will be as shown for Case 1 in Figure 6.18. The reaction R_1 in the figure would be the edge loading for the roof.

If the wall is constructed with ordinary platform framing details, as shown in Figure 6.8, the wall action will be as shown for Case 2 in Figure 6.18. The parapet becomes a separate, freestanding wall on top of the roof, and it must be braced independently for the wind force. The means of bracing depends on the height of the parapet as well as other considerations, such as the finishing materials used, the type of roofing, the means for roof drainage, the type of roof framing system, and so on. The

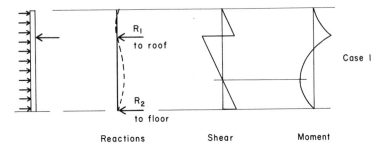

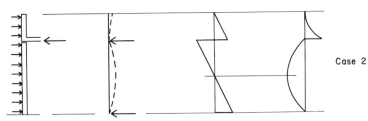

FIGURE 6.18   Optional functioning of the wall and parapet.

illustrations in Figure 6.19 show two possibilities for the bracing of a free-standing parapet.

In the upper figure, a series of diagonal wood braces are nailed to the short studs of the parapet and to a continuous member that is attached to the roof framing through the deck. The diagonal members also function as the framing for the plywood used to achieve the cant at the roof-to-wall intersection. This technique is most useful for relatively short parapets, else the cant would tend to get quite large for a tall parapet.

In the lower illustration in Figure 6.19, a diagonal brace is attached to the back face of the parapet and to the top of the roof structure. The brace may be of metal or wood, and the details for its attachment and the desired spacing of the braces would be critical design factors. The most undesirable feature of this technique is the attachment to the roof, which presents a problem with regard to the roofing. A second undesirable feature is the exposure of the brace and the attachments, which is likely to

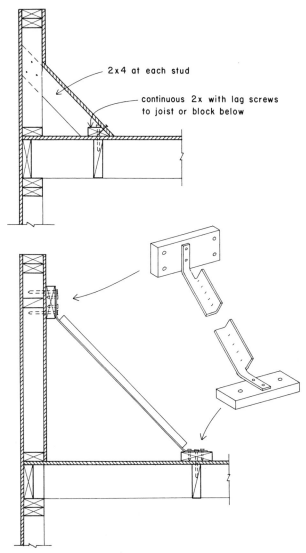

2x4 at each stud

continuous 2x with lag screws
to joist or block below

FIGURE 6.19 Parapet framing details—wood construction.

produce deterioration of the structure. Careful consideration should be given to both of these problems if this technique is used.

If the permanent construction of the roof and ceiling is light, the need for anchorage of the roof against uplift must be considered. For uplift with a flat roof *UBC* Section 2311(c) requires that a force of three-quarters of the horizontal design wind pressure must be resisted. If we assume a total dead load of 12 lb/ft² for the roof and ceiling, the uplift analysis is as follows:

Uplift pressure: $3/4 (20) = 15$ lb/ft²
Net uplift force: $15 - 12 = 3$ lb/ft²
Total uplift force: $(3 \text{ lb/ft}^2)(30 \text{ ft})(50 \text{ ft}) = 4500$ lb

which results in an average uplift force at the roof edge of

$$\frac{4500}{2(30 + 50)} = \frac{4500}{160} = 28 \text{ lb/ft}$$

This force should be easily resisted by the ordinary framing connections at the joint between the roof and walls. However, the connections should be developed with this function in mind.

Vertical and horizontal elements that span across openings must be designed for the horizontal wind pressure on the walls. Headers that span over openings should be designed for the horizontal wind load as well as for the vertical gravity loads. Glazing should be designed for wind pressure using the provisions and tables in Chapter 54 of the *UBC* (see the Appendix) or the recommendations of glazing manufacturers, if they are more stringent. Although the selection of the appropriate thickness is important, the details of the support of the glass edge are also critical and should be designed so that the glass has some freedom of movement.

Because of the exposure of signs as well as the effect on them of gusts, the attachment of signs and other objects to the exterior of the building should be conservatively designed for wind pressure.

For the design considerations for earthquake forces on the building in Figure 6.1 see Example 1 in Chapter 7.

FIGURE 6.20 Building plan (Example 2).

6.2 Example 2

Example 2 is essentially the same as Example 1 with the exception that the details of the front wall exclude the possibility of a shear wall. The building plan is shown in Figure 6.20. There are two possibilities to consider for this type of building. The first is to use the shear walls on the three sides only, in which case the east–west wind must be resisted by the combination of direct load on the rear wall plus torsion on the three-sided structure. As shown in Figure 6.21, the center of stiffness moves to the rear

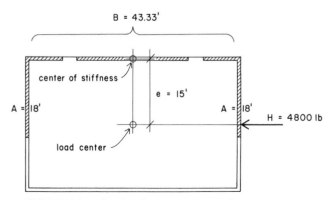

FIGURE 6.21 Torsion due to east–west wind (Example 2).

wall and the torsional moment of inertia, J, is reduced to that provided by the two end shear walls. Referring to the calculations in Table 6.1, this reduces the J for the three-sided building to 22,500 ft^4.

The rear wall must resist the total wind load in direct shear. Using the load of 4800 lb as determined in Example 1, the average stress in the rear wall thus becomes

$$v = \frac{4800}{43.33} = 111 \text{ lb/ft}$$

The roof deck in this case virtually cantilevers from the rear wall. Thus, there is no edge shear stress at the front of the building, and the edge shear at the rear wall is twice that in Example 1. In this example this stress, 96 lb/ft, is still well below a critical value for the roof plywood.

The end walls develop a stress because of the torsional effect of the east–west wind as follows:

$$v = \frac{Hec}{J} = \frac{(4800)(15)(25)}{22,500} = 80 \text{ lb/ft}$$

This stress should be compared with that produced by the north–south wind, which in this example is much higher. Thus, the torsional effect is not critical for the end walls.

The higher value of shear in the rear wall is still not critical for the wall surfacing in this example, but the increase in the sliding and overturn should be considered. If the piers are considered to act individually (Case 1 in Figure 6.14), the loading for the short pier will be as shown in Figure 6.22, and the overturn analysis is as follows:

Lateral load:	(11 lb/ft)(8.67 ft)	=	962 lb
Overturn M:	(962)(11 ft)(1.5 SF)	=	15,873 lb-ft
Dead load M:	(3000)(4.33 ft)	=	12,990
Moment required for hold down:		=	2,883 lb-ft

$$\text{Required } T = \frac{2883}{8.67} = 333 \text{ lb}$$

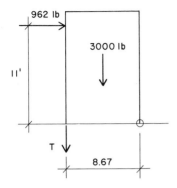

FIGURE 6.22 Overturn analysis for the rear shear wall (Example 2).

This is a quite nominal requirement because the corner framing, the door headers, and the sill bolting all provide additional hold-down resistance. It would be conservative to provide an anchored hold-down device, although it may be well to consider the use of Case 2 in Figure 6.14. The cost of the hold downs would be about the same as that of the blocking and strapping, and the latter would provide for reinforced corners at the openings, as discussed in Example 1.

For the three-sided building, the dimension of the horizontal diaphragm perpendicular to the open side and the ratio of that dimension to the length of the open side are restricted by Section 2514(a) of the *UBC*. For the wood structure the maximum depth of the diaphragm is limited to 25 ft or the width dimension, whichever is smaller, which means that the building in this example is in violation of the code because this dimension is 30 ft. The code does provide that the limit may be increased if "calculations show that the diaphragm deflections can be tolerated." The general problem of deflections, including that of this type, are discussed in Chapter 8.

The alternative to the three-sided box for this building is to use a braced frame or rigid frame for the front wall structure. The design of these types of structures are discussed in Examples 14 and 16.

6.3   Example 3

The building in this example is the same as that in Example 1, except that the roof structure is a light steel frame with a formed sheet steel deck and the walls are of reinforced concrete block. The wind loads on the building and the distribution of loads to the roof deck and end shear walls for wind in the north–south direction will be the same as in Example 1. The differences resulting from the change of construction materials are as follows:

1.   Because of the higher dead weight of the walls, overturn is less likely to be critical. Also, because the doweling of the wall reinforcing into the foundations provides a natural anchorage, additional hold downs are seldom required, even when the dead load resistance is not sufficient.

2.   The rear wall would be designed as Case 2 in Figure 6.14 if the masonry is continuous over the door openings, because the ordinary detailing of the wall reinforcing provides a strong reinforcing of the corners of the openings.

3.   Table No. 24-H of the *UBC* requires that the lateral load for seismic design be increased by 50% when calculating the shear stresses in the wall. This increase is not required for wind, however.

4.   Section 2310 of the *UBC* requires that the roof-to-wall connection be designed for an outward load on the wall of 200 lb/ft or the calculated force, whichever is critical.

5.   For determination of the distribution of loads to the piers in the east–west direction, the pier stiffnesses are determined on the basis of flexural stiffness. This is different from the procedure used for the plywood walls, in which pier stiffnesses were assumed to be proportionate to the wall lengths.

The maximum shear stress of 133 lb/ft, as calculated for Example 1, is quite low for the steel deck. Assuming the maximum deck span to be 6 ft, the deck units will probably be placed in a minimum unit length of 3 spans, or 18 ft. We will therefore assume that the gravity design will be satisfied with a type B deck of 22-gauge steel. For the lateral load design, decisions to be made include the method of attachment of the deck to its steel supports and the means for connecting adjacent deck units at their common edges, called the *side seams*.

Data for the lateral load design of steel decks can be obtained from various sources. In the Appendix we show an excerpt from the *Inryco Lateral Diaphragm Data Manual 20–2* (Ref. 9). The table shown in the Appendix is for type B deck with a deck unit width of 24 in., side seams button-punched at 36-in. centers, and attachment to supports with 3 puddle welds per unit, as shown in the sketch accompanying the table. For our example with the 22-gauge deck on a 6-ft span, the allowable lateral shear is given in the table as 240 lb/ft (labeled "q" in the table.) The F values given in the table (27.87 + 187.0 R for our example) are for determination of the relative flexibility of the diaphragm and calculation of the diaphragm deflection, as discussed in Section 8.1 of Chapter 8.

The determination of pier and total wall stiffnesses for the front and rear walls is illustrated in Figure 6.23. Because of the large opening width and short header height in the front wall, the pier is assumed to be cantilevered from the fixed base. For the rear wall, due to the narrow openings and tall header, the piers are assumed to be fixed at both top and bottom. Stiffnesses for piers for both of these conditions are given in the tables in the *Concrete Masonry Design Manual* (Ref. 7) and are reprinted in the Appendix. The stiffness of individual piers ("R" in the table in Figure 6.23) is determined as a function of the ratio of height to length ("h/d" in the table.) The total stiffness of the rear wall, for comparison to the front wall, is the sum of the R values for the three piers.

For the direct shear effect the distribution to the two walls is made in proportion to the wall stiffnesses, as follows:

Rear wall: $V = (4800)\left(\dfrac{4.722}{4.722 + 0.4928}\right) = 4{,}346 \text{ lb}$

Front wall: $V = (4800)\left(\dfrac{0.4928}{4.722 + 0.4928}\right) = 454 \text{ lb}$

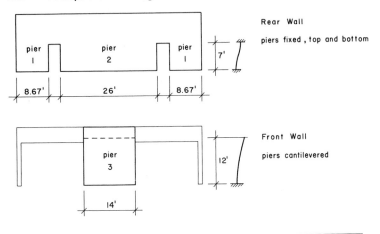

Wall	Pier	h (ft)	d (ft)	h/d	Pier R	No. of piers	ΣR
Rear	1	7	8.67	0.807	0.8485	2	1.697
	2	7	26	0.269	3.025	1	3.025

Total stiffness of rear wall   =   4.722

Front	3	12	14	0.857	0.4928	1	0.4928

Total stiffness of east–west walls   =   5.2148

FIGURE 6.23   Determination of pier and wall stiffnesses (Example 3).

The stress in the front wall pier is thus

$$v = \frac{454}{14} = 32 \text{ lb/ft}$$

In the rear wall the total load of 4346 lb is distributed to the piers in proportion to their stiffnesses, as follows:

Pier 1: $V = \dfrac{0.8485}{4.722}(4346) = 781 \text{ lb}$

$$v = \frac{781}{8.67} = 90 \text{ lb/ft}$$

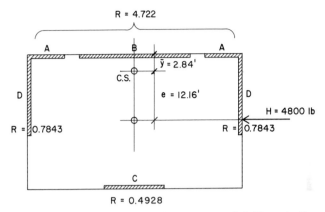

FIGURE 6.24 Torsion caused by east-west wind (Example 3).

Pier 2: $V = \dfrac{3.025}{4.722} = 2784$ lb

$$v = \frac{2794}{26} = 107 \text{ lb/ft}$$

Figure 6.24 illustrates the basis for the determination of the torsional effect of the east–west load. The wall stiffnesses, as determined in Figure 6.23, are used for the calculation of the location of the center of stiffness and the value for the torsional moment of inertia (J). The tabulation for J is shown in Table 6.2.

TABLE 6.2 Torsional Resistance of the Masonry Shear Walls

Wall	Wall Stiffness (see Figure 6.23)	Distance from Center of Stiffness(ft)	No. of Walls	$J = Rd^2$
Front	0.4928	27.16	1	364
Rear	4.722	2.84	1	38
End	(h/d = 0.667)			
h = 12	0.7843	25.0	2	980
d = 18				
	Total torsional moment of intertia (J) =			1382

The location of the center of stiffness can be found by a static moment about the rear wall. Thus

$$\bar{y} = \frac{0.4928(30)}{5.2148} = 2.84 \text{ ft}$$

With J determined in units of wall stiffness, the formula for torsional stress produces wall loads in the same units (pounds per unit of wall stiffness). These values can be multiplied by the actual wall stiffnesses to produce loads for the walls, as follows:

Front wall: $V = \dfrac{Hec}{J} (\text{wall } R)$

$$= \frac{4800(12.16)(27.16)}{1382} (0.4928)$$

$$= 565 \text{ lb}$$

$$v = \frac{565}{14} = 40 \text{ lb/ft}$$

Rear wall: $V = \dfrac{4800(12.16)(2.84)}{1382} (4.722) = 566 \text{ lb}$

End walls: $V = \dfrac{4800(12.16)(25)}{1382} (0.7843) = 828 \text{ lb}$

The torsional stress on the front wall is added to the direct stress previously determined, resulting in a total design stress of

$$v = 32 + 40 = 72 \text{ lb/ft}$$

The torsional stress on the rear wall is ignored because it is opposite in sign to the direct stress. The torsional stress on the end walls is also ignored because it is less than that caused by the north–south wind load.

The stresses determined for both the roof and walls in this example are well below the capacities of ordinary construction, so the design requirements are nominal.

Some possibilities for the roof-to-wall connections are shown in Figure 6.25, where it is assumed that the roof framing consists of open web steel joists spanning the 30-ft dimension and supported

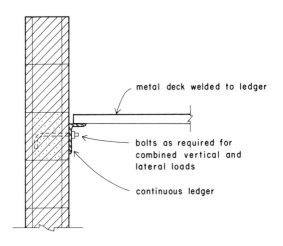

metal deck welded to ledger

bolts as required for
combined vertical and
lateral loads

continuous ledger

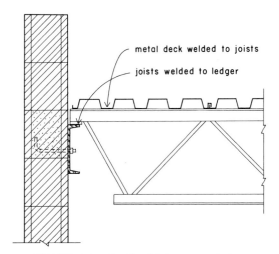

metal deck welded to joists

joists welded to ledger

FIGURE 6.25   Roof/wall joint (Example 3).

at the front and rear by a ledger channel bolted to the wall. At the end walls, the end of the metal deck is supported by a ledger angle that is bolted to the wall. The deck-to-joist, deck-to-ledger, joist-to-ledger, and ledgers-to-walls connections must be adequate to transfer the combinations of gravity and wind loads. The roof-to-wall connection must also be capable of resisting the minimum outward load on the wall of 200 lb/ft, as specified by Section 2310 of the *UBC*.

The masonry walls must also be designed for the direct wind pressure, as was discussed for the design of the studs in Example 1. This loading condition is usually the basis for the determination of the vertical reinforcing in the wall, whereas the horizontal diaphragm shear load is usually the basis for the determination of the horizontal reinforcing. When both of these stress conditions are low, as in this example, the reinforcing is usually established on the basis of the minimum requirements as given in Section 2418(j)3 of the *UBC*.

Because the masonry walls are heavier than the walls in Example 1, overturn is not critical for the individual piers. In any event, end anchorage of the piers is provided by the doweling of the vertical reinforcing at the ends of each pier.

The effect of the gravity load plus the overturning moment on the wall foundation should be investigated, especially if the supporting structure consists of a shallow grade wall. Design for this condition is illustrated in other examples in which the load magnitudes are higher.

6.4 Example 4

As shown in Figure 6.26, this example is similar to Example 1, except for the longer east–west plan dimension and the higher walls. These changes produce higher stresses in the roof diaphragm and higher stresses and overturning effects in the shear walls on the ends of the building. The investigation is limited to these two problems in this example.

The loading of the roof diaphragm by wind on the front of the building is shown in Figure 6.27. For this example, the front wall

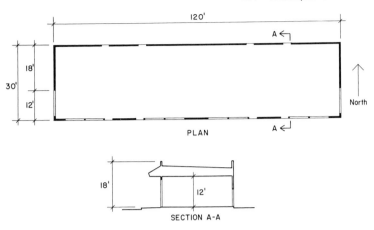

FIGURE 6.26   Building (Example 4).

studs are assumed to cantilever to produce the parapet. Thus, the roof edge loading (R_1 in the figure) is found from a static moment about the bottom of the stud, as follows:

$$R_1 = \frac{(20 \text{ lb/ft})(18 \text{ ft})(9 \text{ ft})}{(15 \text{ ft})} = 216 \text{ lb/ft}$$

which produces the equivalent load zone of 10.8 ft shown in the figure.

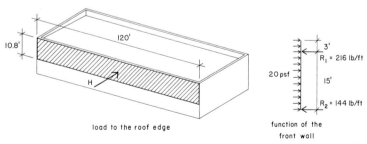

FIGURE 6.27   Front wall function and roof edge loading for north–south wind (Example 4).

The total wind force is thus

$$H = (216 \text{ lb/ft})(120 \text{ ft}) = 25920 \text{ lb}$$

and the maximum shear at the end of the roof deck is

$$v = \frac{25,920}{2(30)} = 432 \text{ lb/ft}$$

Using Douglas fir plywood for the roof deck (see the Appendix for *UBC* Table 25-J), some options are as follows:

1/2 in. Structural II with 2× framing, blocking, and 8d nails at 2½ in. at boundaries and at 4 in. at other edges.

5/8 in. Structural II with 3× framing, blocking, and 10d nails at 4 in. at boundaries and at 6 in. at other edges.

1/2 in. Structural I with 3 × framing, blocking, and 10d nails at 4 in. at boundaries and at 6 in. at other edges.

Because the high stresses occur only near the ends of the building, it is reasonable to consider the possibility of zoning the deck nailing in this case. By using the 1/2-in. Structural II plywood, it is possible to use two fewer nail spacings, as given in the *UBC* table. Thus, the range of nail spacings and corresponding load ratings are the following:

8d at 2½ in. at boundaries, 4 in. at other edges: load = 530 lb/ft.

8d at 4 in. at boundaries, 6 in. at other edges: load = 360 lb/ft.

8d at 6 in. at all edges: load = 270 lb/ft.

In Figure 6.28 these allowable loads are plotted on the graph of wind stress variation in the deck to permit the determination of the areas in which the various nailing spacings are usable. The maximum nailing is seen to be required on only 10 ft at each end of the roof, or only one-sixth of the total roof surface. The actual dimensions of the specified nailing zones may be adjusted slightly to correspond to modules of the roof framing and plywood sheet layouts, so long as the limits of the calculated zone limits are not exceeded.

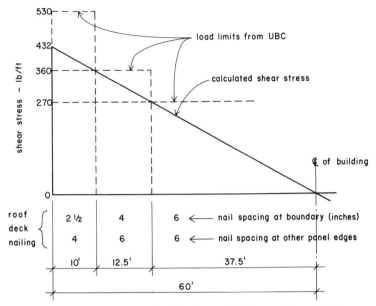

FIGURE 6.28   Zoned nailing for the roof diaphragm (Example 4).

In the end shear wall, the edge load from the roof results in a maximum stress of

$$v = \frac{12,960}{18} = 720 \text{ lb/ft}$$

Because this result is in excess of the maximum capacity of the nailing at the end of the roof diaphragm (530 lb/ft as previously shown), a direct transfer of the total load from the horizontal to the vertical diaphragm is not possible. A collector over the wall opening must be used to push and pull the roof edge load at that location into the shear wall. Using the distributed shear for the end of the roof, as previously calculated, the load in the collector is:

$$T = C = (432 \text{ lb/ft})(12 \text{ ft}) = 5184 \text{ lb}$$

Whatever parts of the roof or wall framing are used for this function, the elements used, as well as any splices in them, must be designed for this force in both tension and compression.

Using Douglas fir plywood for the wall sheathing (see the Appendix for *UBC* Table 25–K), possible options are:

1/2 in. Structural II with 3× boundary framing and 10d nails at 2 in. at all edges: load = 770 lb/ft.

3/8 in. Structural I with 8 d nails at 2 in. at all edges: load = 610(1.2) = 732 lb/ft. (See footnote No. 3 in the *UBC* Table of the Appendix.)

1/2 in. Structural I with 3× boundary framing and 10d nails at 2½ in. at all edges: load = 770 lb/ft.

And if it is possible to have plywood on both sides of the wall, use:

3/8 in. Structural II, both sides, with 8d nails at 4 in. at all edges: load = 2(320)(1.2) = 768 lb/ft.

Although technically permitted by the code, none of the foregoing is desirable. The closely spaced, heavy nailing is likely to cause some splitting of framing members. Placing plywood on both sides of the wall can entail some difficulty in the framing details and in the installation of wiring or other elements within the wall cavity. The heavy shear load on the wall is also likely to be a problem in terms of stress on the wall boundary framing, a need for heavy anchorage, or an excessive load on the foundations. Deflection may also be a problem, especially if the wall is relatively tall with respect to its plan length. With load magnitudes of this range, all these situations should be carefully investigated.

Given loading and dimensions of the wall as shown in Figure 6.29, the overturn analysis is as follows:

Lateral load:	Total $H/2$ = 25,920/2 =	12,960 lb
Overturn M:	12,960(15 ft)(1.5 SF) =	291,600 lb-ft
Dead load M:	(8000)(9) =	72,000
Moment required for hold down:		219,600 lb-ft

Required $T = \dfrac{219,600}{18} = 12,200$ lb

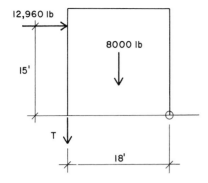

FIGURE 6.29 Overturn analysis for the end wall (Example 4).

This is a considerable tie-down force that requires a very heavy anchorage device, heavy end framing in the wall, and a strong foundation.

Development of sliding resistance for the load of 12,960 lb requires heavy bolting of the wall sill to the foundation. In order to keep the number of bolts to a reasonable limit they must be 3/4 in. or larger in size.

For the roof the edge loading of 216 lb/ft and the diaphragm span of 120 ft and depth of 30 ft result in a chord force of:

$$T = C = \frac{(216)(120)^2}{(8)(30)} = 12,960 \text{ lb}$$

Although this load may be developed with the ordinary framing, the length of the building requires several splices, each of which needs considerable bolting to develop this chord force.

A final consideration is the horizontal deflection of the roof diaphragm at the center of the 120-ft span. The diaphragm depth-to-span ratio of 4 to 1 is just at the limit permitted by the *UBC* (see *UBC* Table 25–I in the Appendix). Even though the diaphragm is within the code limit, the actual dimension of the deflection should be determined and its possible effect on interior partitions considered.

We abandon this example without attempting to solve all its problems because we actually consider it to be a poor solution. It was presented to show what bad planning can produce in the way of difficult situations and not as an illustration of proper design.

FIGURE 6.30 Building plan (Example 5).

6.5 Example 5

As shown in Figure 6.30, this example consists of the same building as used in Example 4 with the addition of an interior shear wall. With the same roof edge loading of 216 lb/ft, as in Example 4, the analysis of the roof diaphragm is shown in Figure 6.31. In this analysis the roof diaphragm is assumed to be

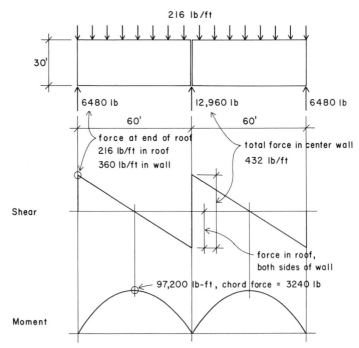

FIGURE 6.31 Analysis of the roof diaphragm assuming peripheral distribution (Example 5).

sufficiently flexible to justify an assumption of distribution of load to the vertical elements on a peripheral basis, which is consistent with the usual practice for a wood diaphragm with the depth-to-span ratio in the example. With this assumption, half of the total wind load is taken by the center wall, and the load on the end walls is reduced to half that in the previous example.

The maximum stress in the roof diaphragm drops to 216 lb/ft, which occurs at the ends and at both sides of the center wall. This is less than the lowest rated capacity for 1/2 in. plywood with edge blocking and minimum nailing as given in *UBC* Table 25–J (see the Appendix.) Nail zoning is therefore not a consideration, unless it is acceptable to use an unblocked diaphragm for some portion of the roof.

The shear stress in the end wall drops to 360 lb/ft. This is still a significant stress, but it can be achieved with 3/8 in. Structural II plywood with 8d nails at 4 in. at all edges, which is quite reasonable. The overturn and tie-down requirement are also considerably reduced, as shown in Figure 6.32 and the following calculation.

Lateral load: $H/4 = 6480$ lb (see Figure 6.31)
Overturn M: $(6480)(15)(1.5) = 145,800$ lb-ft
Dead load M: $(8000)(9)   =   \underline{72,000}$

Moment required for hold down: 73,800 lb-ft

Required $T = \dfrac{73,800}{18} = 4100$ lb

which is within the capacity of a reasonably sized hold-down device.

The center shear wall must carry a considerable force—the same as that on the end shear walls in Example 4. However, the wall is longer in plan, which results in a lower shear stress and less overturn effect. The design of this wall must include the full consideration of its use in the building. Possible issues are its use for fire separation, acoustic separation, and load bearing for the roof structure. For our analysis we consider it to be a single stud wall and to serve as a bearing wall for the roof framing. If it were not a bearing wall, the overturn would be slightly higher than that calculated.

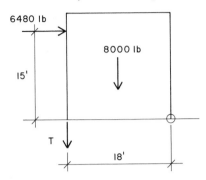

FIGURE 6.32 Overturn analysis of the end shear wall (Example 5).

For the shear load of 12,960 lb, the unit shear stress in the wall is

$$v = \frac{12,960}{30} = 432 \text{ lb/ft}$$

Two materials we might possibly use are:

3/8 in. Structural II plywood on both sides with 8d nails at 6 in. at all edges: load = 2(220)(1.2) = 528 lb/ft.

1/2 in. Structural II plywood on one side with 10d nails at 4 in. at all edges: load = 460 lb/ft.

On the basis of the surfacing alone, the more economical choice would be the single surface of 1/2 in. plywood. However, for acoustic separation the wall may be built as a double stud wall, in which case the two surfaces would probably be more logical.

With the loads assumed as shown in Figure 6.33, the overturn analysis for the center wall is as follows:

Lateral load: $H/2 =$ 12,960 lb (See Figure 6.31)
Overturn M: (12,960)(15)(1.5) = 291,600 lb-ft
Dead load M: (20,000)(15)   = <u>300,000</u> lb-ft

Moment required for hold down: 0

If the wall is not a bearing wall, there may be some tie-down requirement. However, the connection of this wall to the walls at

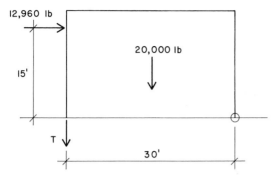

FIGURE 6.33 Overturn analysis of the interior shear wall (Example 5).

the front and rear of the building would probably be adequate for this function.

If sliding friction is ignored and the entire horizontal force is taken on the sill bolts, the choice is whether to use a lot of small bolts or a few large ones. This is a matter of individual preference by designers and builders. If the maximum code spacing of 6 ft is used, the two walls will have a minimum of 4 bolts in the end wall and 6 in the center wall. Thus, the minimum required capacity for this minimum number of bolts is as follows:

End wall: $\text{Load} = \dfrac{6480}{4} = 1620$ lb/bolt

Center wall: $\text{Load} = \dfrac{12,960}{6} = 2160$ lb/bolt

If the bolt size required for these loads is not excessive, the minimum number of bolts may be used—which is the usual preference of both the concrete and framing contractors.

A construction detail that must be developed for this example is the connection between the roof diaphragm and the center shear wall. This connection must transfer the total force of 432 lb/ft from the roof to the wall. Because there are several options for both the roof and wall construction, the potential variations for this connection become quite numerous. Figure 6.34 illustrates some of the possibilities, based on the assumption that the wall is

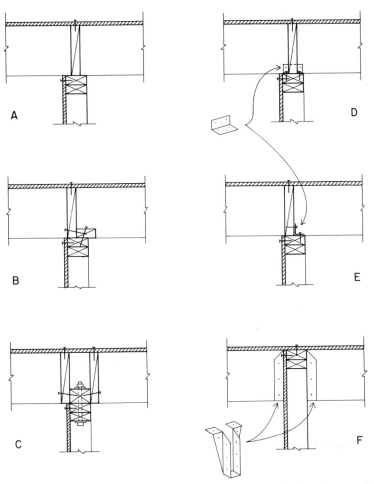

FIGURE 6.34 Options for the roof/wall joint at the interior shear wall (Example 5).

a single stud wall and that the roof framing consists of joists perpendicular to and supported by the wall. There are three basic structural functions to be considered for this situation: the vertical gravity load transfer, the hold down against wind uplift, and the shear transfer of wind force parallel to the wall.

Detail "A" in Figure 6.34 shows the ordinary construction used if the gravity load alone is considered. The joists are either butted end to end, or they are lapped on top of the wall and are toe nailed to the top plate. Vertical blocking is usually provided between the joists for their stability as well as to provide a nailer for the plywood edges perpendicular to the joists. This joint provides only minor resistance to uplift (relying on withdrawal of the toenails) and virtually no capacity for transfer of the wind shear.

In detail "B" of Figure 6.34, a second, horizontal block is added to facilitate the transfer of wind shear. The vertical block is nailed to the horizontal block, and the horizontal block is nailed to the top plate. With the roof deck nailed to the top of the vertical block, the transfer is achieved from the roof deck to the top plate of the wall. With the wall surfacing nailed to the top plate, the transfer from horizontal to vertical diaphragms is then complete.

It should be noted that, although the stress in the roof deck is only 216 lb/ft at this location (see Figure 6.31), the total load transfer to the wall is twice this, resulting from a delivery of load to both sides of the wall by the deck. Thus, all the nailing shown in detail "B" of Figure 6.34—the deck to the vertical block, the blocks to one another, and the block to the top plate—must be designed for the load transfer of 432 lb/ft. If a roof plywood panel edge occurs at this point, the nailing will be adequate, because there will actually be two edge nailings at the joint. If this is an interior support point for a plywood panel, the usual minimum nailing with nails at 12 in. centers will not be sufficient, and a nailing must be specified that is capable of the load transfer. It is questionable whether the nailing of the blocks for this load magnitude could be achieved without splitting the blocks, so this option is probably not the best for the example.

A variation on detail "B" is shown in detail "C" of Figure 6.34. A second vertical block is added, and the horizontal block is bolted to the top plate. This results in an extended range of load capacity because there are now two rows of nails and the bolts are much stronger than nails in lateral load resistance. However, with fairly closely spaced joists, this detail would require a considerable number of bolts through the top plates.

Another approach to this connection is shown in details "D" and "E" of Figure 6.34, in which metal framing anchors are used for the attachment to the top plates. In detail "D" the anchors are attached to the joists, and in detail "E" they are attached to the blocks. Assuming that the anchor devices used have a rated load capacity adequate for the load transfer, either of these options is acceptable for the wind shear function. Detail "D" is slightly better for wind uplift because there is more attachment between the roof deck and the joists than there is between the deck and the blocks.

A third technique for this connection is shown in detail "F" of Figure 6.34, in which the top plate of the wall is raised to the level of the roof deck. This procedure allows for the simplest transfer of the wind shear because both the roof deck and the wall sheathing are directly nailed to the top plate. The joists are supported by saddle-type metal hangers hung from the top plate. One problem with this detail is that the upper panels of the wall sheathing must be installed before the roof framing can be placed, which is not the usual sequence of the construction.

We do not attempt to judge which of these, or of other possible options, is the best solution for this connection. From the viewpoint of structural design, anything that "works" is all right. In real situations there are many issues to consider in addition to the necessary structural functions. Thus, the influence of roof drainage, wall surface finishes, ceiling construction, ductwork installation, and so on may provide the deciding factors for choice between viable alternatives.

6.6 Example 6

This example consists of the same building plan and form as used in Example 5. In this case the roof is made of steel deck on steel joists, and the shear walls are made of reinforced concrete masonry. The steel deck produces a diaphragm slightly stiffer than that produced by plywood. Unless there is a structural concrete fill on top of the steel deck, however, the diaphragm is still probably flexible enough to justify an assumption of distribution to the shear walls on a peripheral basis, rather than a pier stiffness basis.

Determined on a peripheral basis, the distribution of load from a north–south wind would be the same as shown in Figure 6.31 for Example 5. To illustrate the process we will perform a distribution on the basis of pier stiffness, assuming the walls to be cantilevered from fixed bases. The calculations for this analysis are shown in Table 6.3. The stiffness values for the piers (R in the table) are taken from the tables in the *Concrete Masonry Design Manual* (Ref. 7; see also the Appendix.) The total wind load of 25,920 lb is distributed to the shear walls in proportion to their stiffnesses, as follows:

On the end wall: $V = \dfrac{0.519}{2.288} (25{,}920) = 5880 \text{ lb}$

On the center wall: $V = \dfrac{1.250}{2.288} (25{,}920) = 14160 \text{ lb}$

Assuming these results to be the reactions for the roof diaphragm, the distribution of shear and moment in the roof will be as shown in Figure 6.35. Comparison with the values in Figure 6.31 show that the difference in results between the two assumptions for distribution in this example is minor. Because of this minor difference, we use the diaphragm stresses calculated in the previous example for discussion of this structure. From the pre-

TABLE 6.3   Determination of Wall Stiffness for Distribution of North–South Wind Load.

Wall	Height $-H$ (in feet)	Length $-D$ (in feet)	H/D	Stiffness (R)	No. of walls	Total R
End	15	18	0.833	0.519	2	1.038
Center	15	30	0.50	1.250	1	1.250
	Total stiffness of north–south walls:					2.288

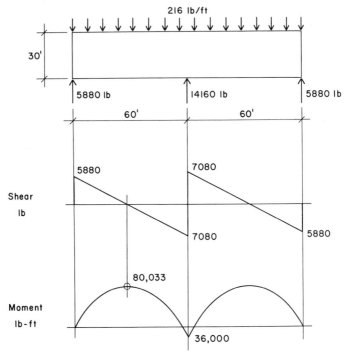

FIGURE 6.35   Analysis of the roof diaphragm assuming rigid diaphragm action (Example 6).

vious calculations, these stresses are 216 lb/ft in the roof, 360 lb/ft in the end shear walls, and 432 lb/ft in the center shear wall.

The stress in the roof deck is well below the capacity of typical light gauge type A or B steel deck (see the Appendix). The stress in the walls is also well below the capacity for reinforced masonry. If reinforced masonry is used, the reinforcing should be selected on the basis of minimum code requirements. The end walls must also be designed for the combination of gravity loads and the effect of direct wind pressure, which create a situation of axial compression plus bending on the walls.

Because of the low stresses, it is possible to consider the use of unreinforced masonry walls if they are permitted by the local building code. If unreinforced walls are used, the wall thickness is limited by the maximum slenderness ratio (unbraced height-to-wall thickness) as limited by the building code. Section 2419(b)2 of the *UBC* limits this ratio to 18 for hollow-unit masonry. For the 15-ft-high wall in this example, this would require a minimum thickness of $15(12)/18 = 10$ in.

If a block of 12-in. nominal thickness is used with an average net solid cross section of 45%, the actual net stress on the center wall is:

$$v = \frac{432 \text{ lb/ft}}{(12)(11.625)(0.45)} = 6.9 \text{ psi}$$

From Table 24–B of the *UBC* we find that the use of type S mortar would allow a maximum stress of $(1.33)(6) = 9$ psi. It should be noted, however, that the *UBC* permits the use of unreinforced masonry only in seismic risk zone 1 (see *UBC* Section 2312(j)2.B).

Overturn of the shear walls in this example is much less critical because of their larger dead weight. If the walls are reinforced, there will be some vertical reinforcing at their ends that will be dowelled to the foundation, making end anchorage a natural attribute of the construction.

Because of the low range of stresses in this example, we illustrate the design of masonry shear walls more thoroughly in other examples.

The situations for connection between the roof and exterior walls in this example are similar to those that were discussed and illustrated for Example 3. At the center wall there are a number of possible situations, depending on the type of roof framing and the direction of the joists with respect to the wall. The details in Figure 6.36 illustrate a situation in which the wall serves as a bearing wall with rolled steel joists supported in an end bearing on top of the wall. The ends of the steel joists rest on a steel plate that is bolted to the top of the wall. The wind load must be transferred from the steel deck to the wall. The total mechanism of this transfer is as follows:

1. The deck is welded to the top of the joists.
2. The joists are stud-bolted to the plate.
3. The plate is bolted to the concrete-filled top course of blocks in the wall with preset anchor bolts.

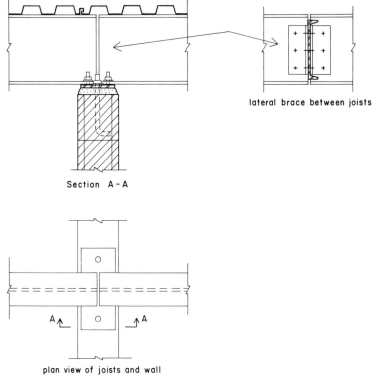

lateral brace between joists

Section A-A

plan view of joists and wall

FIGURE 6.36   Roof/wall joint at the interior shear wall (Example 6).

Lateral bracing of the joists is accomplished by the channel sections shown in Figure 6-36. These sections work in a manner similar to the blocking in the wood structure (see Figure 6.34).

An alternative to this construction is shown in Figure 6.37. In this case a continuous steel-rolled section is bolted to the top of the wall, and the wind load transfer bypasses the joists entirely to provide a much stronger and stiffer connection.

6.7   Example 7

In the previous examples the roof edge occurred at the exterior walls, with the walls extended above the roof level to form a

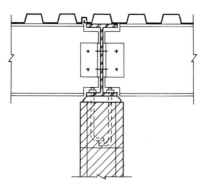

FIGURE 6.37 Optional roof/wall joint at the interior shear wall (Example 6).

parapet. In this example the roof structure is extended to form a cantilevered overhang, as shown in the building section in Figure 6.38. Several possible framing systems could be used for this structure. The partial framing plan in Figure 6.39 shows a system consisting of wood elements. The basic system employs glue-laminated girders on 16-ft centers. These girders support purlins on 8-ft centers, which in turn support joists on 2-ft centers. This system permits the use of standard 4-ft-by-8-ft plywood panels with a minimum necessity for added edge blocking.

In order to form the cantilevered corner, a diagonal framing member is used and the joist direction is changed to provide the cantilever at the building ends. As shown in Figure 6.40, both the girders and the joists are cantilevered at the front and rear walls. There are three problems to consider at this location, as follows:

1. How can one support the cantilevered joists and the facia and soffit construction? One possibility is to provide a purlin at the outer end of the cantilevered girder and to have the joists span in simple span action between this member and the first regular purlin (purlin 2 in the illustration.) Another option is to extend the wall up to the underside of the joists and to let both the girders and joists cantilever from the wall.

2. What can be used for the edge chord for the roof diaphragm? One possibility is the edge purlin (purlin 1 in the illustration). Another is to use the continuous top plates in

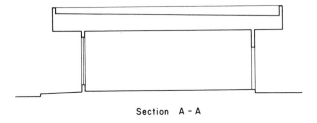

Section   A - A

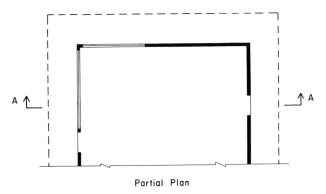

Partial Plan

FIGURE 6.38   Building (Example 7).

the wall. However, in the illustration these plates are shown to occur below the girders, which places them a considerable distance from the roof deck. In order for the plates to be used effectively, they would have to occur at the underside of the deck or the joists, which would require that the girder be notched or have its top dropped to a position below the plates.

3. How can the roof edge shear load be gotten into the wall? In the illustration this is achieved by extending the wall up between the girders and providing blocking between the joists. Nails "B," "C," and "D" in the illustration are used to get the load from the roof plywood to the wall plywood.

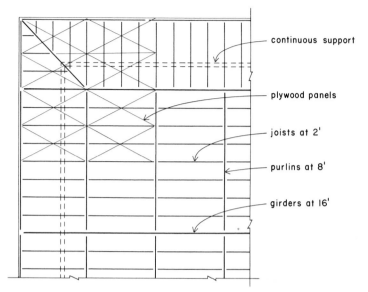

- continuous support
- plywood panels
- joists at 2'
- purlins at 8'
- girders at 16'

FIGURE 6.39   Partial framing plan for the wood roof structure (Example 7).

There are, of course, many considerations that may influence the development of these details, such as the actual magnitude of the forces to be transferred, the size of the girders, the means for providing roof drainage, and so on.

At the building ends, the situation is somewhat simpler because the girders do not cantilever at this point. As shown in Figure 6.41, this makes it possible to place the continuous top plates of the wall at the underside of the joists. In this position they can function to provide support for the joists, a diaphragm chord for the roof, and a collector for the top of the wall.

If the edge purlin (purlin 1 in Figure 6.40) is used as a chord, the depth of the roof diaphragm for calculation of the chord force would be the full width of the roof. However, if the top plates of the wall are used as collectors at the building ends, the maximum stress in the roof deck would be determined using only the wall length as the diaphragm depth, because this is the distance over which the load transfer occurs.

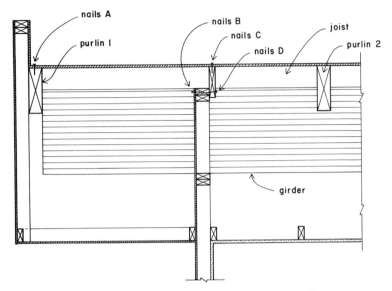

FIGURE 6.40   Section at the front wall (Example 7).

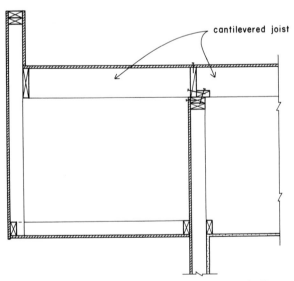

FIGURE 6.41   Section at the end wall (Example 7).

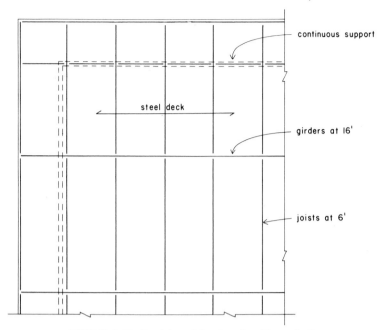

continuous support

steel deck

girders at 16'

joists at 6'

FIGURE 6.42   Partial roof framing plan (Example 8).

6.8   Example 8

This example is similar to Example 7, except that the roof structure is of steel and the walls are of masonry. Figure 6.42 shows a partial framing plan for the roof in which the basic system consists of a series of steel girders on 16-ft centers, which support short steel joists at 6-ft centers. The steel roof deck spans between the joists. The corner cantilever could be developed in a manner similar to that used for the wood structure as shown in Figure 6.39. However, the diagonal member has been left out in this scheme, and the corner is supported through the combined cantilever of the first interior joist and the steel facia framing member.

The same basic problems that were discussed for Example 7 must also be solved for this structure. The steel facia member, as

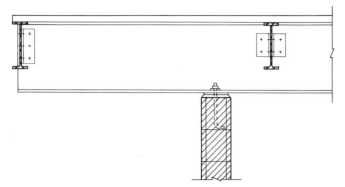

FIGURE 6.43   Section at the front wall (Example 8).

shown in Figure 6.43, would be used for the diaphragm chord, with the standard angle framing connections probably providing adequately for the chord continuity. The combined effects of the facia and the first interior joist, both being made continuous through their connections to the girders, would probably adequately provide the collector function for the front wall.

With the framing as shown in Figure 6.43, the east–west wind load must be transferred from the deck through the joists and into the girders as a lateral load. This load is then transferred into the wall through the anchor bolts in the bottom flanges of the girders, which tends to cause the girders to roll over laterally and is resisted principally by the moment capacity of the joist connections. If the joists are relatively shallow in depth, this procedure may not be an adequate means for bracing the girders. One solution to this problem would be to use a row of X-bracing along the wall, as shown in Figure 6.44. Another solution would be to place the first joist over the wall and make it slightly deeper with a connection offering more moment resistance.

This rollover problem also occurs at the end walls, requiring the addition of a framing member between the joists, as shown in Figure 6.45. This member can also serve the collector function if this function is required. If the collector requirement is not a factor, it would also be possible to use X-bracing for the rollover resistance, as was shown for the girder in Figure 6.44.

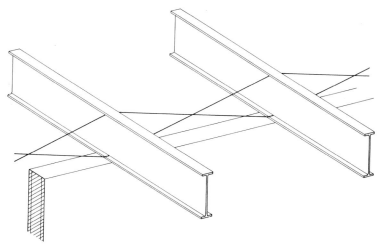

FIGURE 6.44   Optional lateral bracing of the roof beams at the front wall (Example 8).

There are many other ways to achieve the structure for the roof and many other detail functions to consider, such as roof drainage, framing of the facia and soffit for the overhang, spanning over window openings, and so on. Thus, the possible details for achieving the basic tasks of providing the diaphragm chord, the collector, and the load transfers are quite numerous.

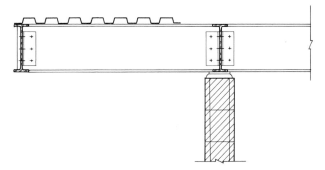

FIGURE 6.45   Section at the end wall (Example 8).

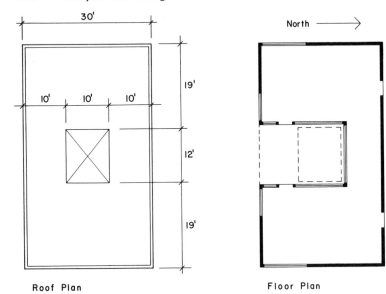

FIGURE 6.46   Building plans (Example 9).

6.9   Example 9

This example is similar to Example 1, except that there is a large hole in the roof for a skylight, as shown in the building plan and roof plan in Figure 6.46. With no interior vertical bracing elements, the roof diaphragm must still span from end to end of the building and sustain the shear and bending stresses that result. The shear and moment diagrams for the north–south loading are shown in Figure 6.47 (see Figure 6.3, Example 1.) As shown in the diagram, the critical shear at the edge of the hole is 960 lb, which must be taken by the net diaphragm width of 20 ft, resulting in a stress of

$$v = \frac{960}{20} = 48 \text{ lb/ft}$$

which is quite low, so that diaphragm shear is not critical.

The critical stress condition is actually that which occurs at the corners of the hole. For this stress it is important that the edge

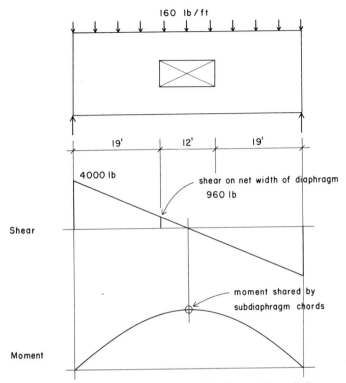

FIGURE 6.47 Analysis of the roof diaphragm (Example 9).

framing on all sides be made continuous into the adjacent diaphragm, as was done at the corners of the door openings in the
rear shear wall in Example 1 (see Figure 6.17.) The means for
achieving this depends somewhat on the details of the roof framing system. The partial framing plan in Figure 6.48 shows one
possibility in which the east and west edges of the hole are
bounded by main framing members that span the building width.
These members provide for the necessary continuity in the
north–south direction at the corner. Continuity in the east–west
direction is provided by placing framing members in the adjacent
bays in line with the edge framing on the north and south edges of
the hole. These members are strapped to the edge framing to

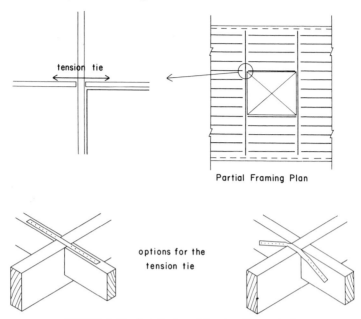

FIGURE 6.48   Detail of roof framing (Example 9).

achieve continuity in a way similar to that for the shear wall in Example 1. To determine the force required for the straps it is assumed that the roof diaphragm at this location consists of two subdiaphragms, each 10 ft wide. The total moment of 50,000 lb-ft must be shared by these two subdiaphragms. On this basis, the chord forces and the stress in the strap are

$$T = C = \frac{1/2 \ (50,000)}{10} = 2500 \ lb$$

As shown in the sketches in Figure 6.48, the tension tie at the corner may be developed with a single tie strap on the top of the members or with a twisted tie that crosses the main beam and is nailed to the sides of the two separate members. If the force is sufficient and if the edge framing is the same depth as the main beams, it is possible to use a flat strap on both the top and the bottom.

Although it is most likely not critical in this example, the effect of openings on the diaphragm deflection should also be considered. Calculation of this effect is not possible, but it should be recognized that there will be some loss of stiffness resulting in a deflection slightly more than that calculated for a diaphragm with no hole.

6.10   Example 10

This example consists of a simple two-story box, as shown in the plan and section in Figure 6.49. The wind load in the north–south

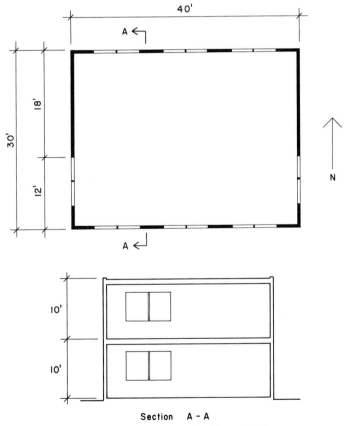

Section   A - A

FIGURE 6.49   Building plan (Example 10).

direction is resisted by the two end shear walls. As shown in Figure 6.50, the loading on these walls consists of the two end reactions from the roof and second floor diaphragms. Using a wind pressure of 25 psf, the magnitudes of these loads are:

$$H_1 = 1/2 \ (25\text{psf})(5 \ \text{ft})(40 \ \text{ft}) = 2500 \ \text{lb}$$

$$H_2 = 1/2 \ (25)(10)(40) = 5000 \ \text{lb}$$

The wall loading condition, including the dead loads, is as shown in Figure 6.51, and the shear stress and overturn analyses are as follows:

Second-story wall:

$$v = \frac{2500}{18} = 139 \ \text{lb/ft}$$

Overturn M: $(2500)(10)(1.5 \ SF) = 37{,}500 \ \text{lb-ft}$
Dead load M: $(5000)(9) \qquad\quad = 45{,}000 \ \text{lb-ft}$

(indicating no hold down requirement)

First-story wall:

$$v = \frac{7500}{18} = 417 \ \text{lb/ft}$$

Overturn M: $(2500)(20)(1.5 \ SF) = 75{,}000 \ \text{lb-ft}$
          $(5000)(10)(1.5 \ SF) = \underline{\ \ 75{,}000\ \ }$
          Total $= 150{,}000 \ \text{lb-ft}$
Dead load M: $(10{,}000)(9) \qquad = \underline{\ \ 90{,}000\ \ }$
Moment required for hold down: $60{,}000 \ \text{lb-ft}$

$$\text{Required } T = \frac{60{,}000}{18} = 3333 \ \text{lb}$$

Assuming the walls to be of plywood, it would be necessary to use the minimum code required nailing at the second story and fairly heavy nailing at the first story. The tie-down force is significant, but it is within the capacity of a reasonably sized tie-down device.

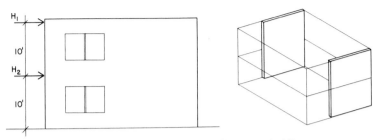

FIGURE 6.50   End shear walls (Example 10).

Stresses in the horizontal diaphragms are not critical unless there are large openings that reduce their effective width. Figure 6.52 shows a plan for the second floor with a large stair opening next to one end wall. The stair opening reduces the width of the second floor diaphragm to 18 ft at the end, which results in a maximum stress of

$$v = \frac{5000}{18} = 278 \text{ lb/ft}$$

This is still not very high for the deck, which would probably be a minimum of 1/2 in. plywood. As shown in Figure 6.53, it would be desirable to reinforce the corners of the opening with straps and blocking, as was done for the roof opening in Example 9.

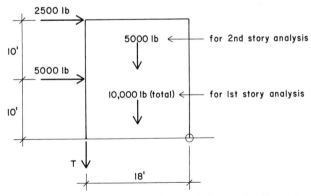

FIGURE 6.51   Overturn analysis of the end shear walls (Example 10).

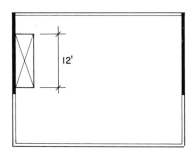

FIGURE 6.52 Second floor plan (Example 10).

The location of the opening creates a problem in achieving the transfer of force from the floor to the end wall. With the framing as shown in Figure 6.54, the header built into the wall may be used to effect the transfer if the header is continuous or is adequately spliced. Care should be taken to assure that the plywood on the end wall is nailed to this header with nails having the capacity to achieve the transfer from the second floor diaphragm.

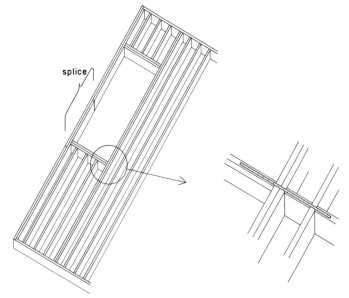

FIGURE 6.53   Framing at the stair opening (Example 10).

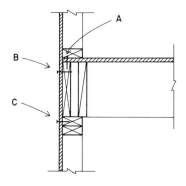

FIGURE 6.54 Floor/wall joint (Example 10).

Figure 6.54 shows a typical detail for the construction of the end wall at the second floor. Nails "A" in the figure are used to transfer the load from the floor plywood to the header. However, the load should preferably be delivered to the top plates of the first-story wall because they more fully provide continuity at the top edge of the lower shear wall. Thus, both nails "B" and "C" in the figure should be used to achieve the complete transfer from the second-floor deck to the top plates of the lower wall. Other means can be used to achieve this result, but for the detail as shown this nailing seems to be the simplest solution.

Although the previous overturn analysis indicated no requirement for a tie down for the second-story wall, we show in Figure 6.55 two typical means for achieving such a tie in light wood framing. Actually, the continuity of the plywood on the wall surfaces at the corner constitutes a considerable tie effect. However, the ties shown, or others, are usually provided when a tie-down force is calculated to be required.

6.11 Example 11

This example consists of the same building as that used in Example 10 except that the construction uses steel framing for the roof and second floor and reinforced masonry for the shear walls. A principal difference effected by this change in materials is that there is considerably more dead load in the shear wall, primarily because of the difference in the weight of the wall itself. It is likely

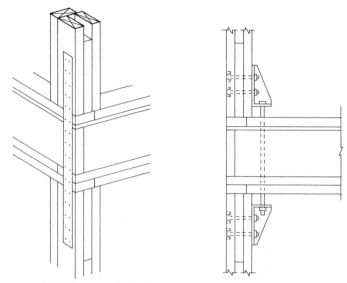

FIGURE 6.55   Hold downs at second floor (Example 10).

that the masonry wall will be at least three times as heavy as the wood stud wall. Assuming the masonry wall to be 40 lb/ft² heavier, the added dead load on the second-story wall will be

$$W = (40)(18)(10) = 7200 \text{ lb, say } 7000 \text{ lb}$$

and the total added load on the first story wall will be

$$W = 7000 + (40)(18)(10), \text{ say } 14,000 \text{ lb}$$

With these loads the total loading of the shear wall will be as shown in Figure 6.56. (For comparison see Figure 6.51 for the loading on the wood stud wall.) A review of the overturn analysis for the wood stud wall should indicate that the added dead load will eliminate the need for a tie down at the first story, although as previously discussed, the masonry construction provides a natural tie for the ends of the shear walls.

Because the stresses are quite low in the shear walls in this example, we will not develop it further. However, the same structure is analyzed in Example 4 in Chapter 7 for a comparison

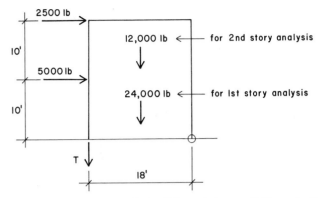

FIGURE 6.56   Overturn analysis of the end shear wall (Example 11).

of the wind and seismic effects. Because the seismic load is critical, the structure is discussed more thoroughly in that example.

6.12   Example 12

As shown in Figure 6.57, this example consists of a three-story building in which the vertical elements of the laterally resistive system for wind in the north–south direction are two interior shear walls. The discussion is limited to the design of these two walls and their foundations.

Assuming a base wind design pressure of 30 psf from the *UBC* wind map, the wind loading is as shown in Figure 6.58. From *UBC* Table 23–F (see the Appendix) the pressure is 25 psf for the first 30 ft of height and 30 psf for the next 20 ft. Determination of the values for H_1, H_2, and H_3, as shown in the illustration, is as follows:

$$H_1 = (30 \text{ psf})(120 \times 7.5) = 27,000 \text{ lb}$$
$$H_2 = (30)(120 \times 12.5) + (25)(120 \times 2.5)$$
$$= 45,000 + 7,500 = 52,500 \text{ lb}$$
$$H_3 = (25)(120 \times 17.5) = 52,500 \text{ lb}$$

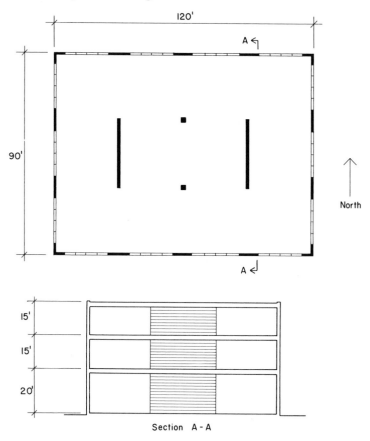

Section A - A

FIGURE 6.57   Building (Example 12).

These loads, together with the assumed dead loads, are shown applied to the three-story-high shear wall in Figure 6.59. It is assumed that the building symmetry permits the total wind load to be divided equally between the two walls.

The magnitude of the wind load precludes the possibility of a plywood diaphragm for the lower portions of the wall. We thus assume a wall of reinforced masonry with hollow concrete units with f'_m of 1500 psi, and the dead loads are based on this con-

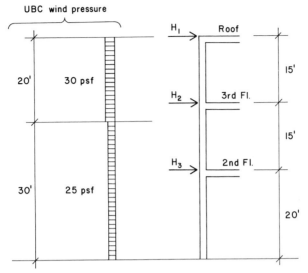

FIGURE 6.58 North–south wind loads (Example 12).

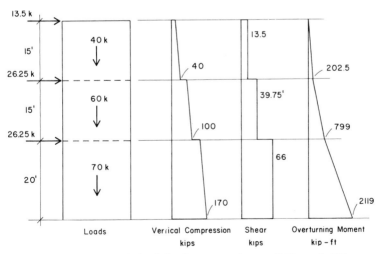

FIGURE 6.59 Analysis of the three-story shear wall (Example 12).

175

struction. We must investigate each story of the wall separately, but we should also consider the vertical continuity of the wall and the problems of the intersections with roof and floor construction.

Investigation of the third-story wall:

Shear stress: $\dfrac{13,500}{30} = 450$ lb/ft

Overturning M: $13.5 \times 15 = 202.5$ k-ft

Dead load M: $40 \times 15 = 600$ k-ft

Safety factor: $\dfrac{600}{202.5} = 2.96$

(No tie down is required)

End force in wall: $\dfrac{202.5}{28.67} = 7.06$ k (plus or minus gravity)

This shear stress and end force are within the capacity range of a plywood wall, although some tie down may be required because the dead load assumes a masonry wall. It is possible if the building design, fire code, and construction details otherwise permit it, that the third-story wall can be built of plywood and the lower stories be built of masonry or concrete.

Investigation of the second-story wall:

Shear stress: $\dfrac{39,750}{30} = 1325$ lb/ft

Overturning M: $13.5 \times 30 = 405$ k-ft
$+ \ 26.25 \times 15 = \underline{394}$
$\qquad$ Total $\qquad = 799$ k-ft
Dead load M: $\quad 100 \times 15 = 1500$ k-ft

Safety factor: $\dfrac{1500}{799} = 1.88$

(No tie down is required)

End force in wall: $\dfrac{799}{28.67} = 27.9$ k (plus or minus gravity)

Assuming an 8-in., 45% solid unit, the shear stress on the net wall section will be

$$v = \frac{1325}{12 \times 7.625 \times 0.45} = 32 \text{ psi}$$

The *UBC* Table 24–H (see the Appendix) gives the minimum usable shear stress as $1.33(35) = 47$ psi. Because the gravity load stress is also quite low, assuming a bearing wall function, the general wall reinforcing can be minimal. A critical design consideration is for the ends of the walls, which must be designed for the wall end forces plus any gravity loads concentrated on the ends by the roof or floor framing.

Figure 6.60 shows a detail of an end column consisting of an 8-in. by 16-in. block, filled with concrete and reinforced as a masonry column. We assume the following design conditions for this column:

Floor dead plus live load = 60 k

Total gravity load plus wind load = $60 + 27.9 = 87.9$ k

Use $f'_m = 1500$ psi, $f_y = 50$ ksi

Design criteria: *UBC*, Section 2418(k)

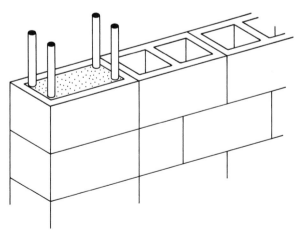

FIGURE 6.60   Shear wall end detail (Example 12).

The *UBC* requires a minimum column dimension of 12 in. and a maximum slenderness ratio of 20. An exception permits an 8 in. column but reduces the allowable axial load to half that for a larger column. If we retain the 8-in. dimension, the maximum unsupported height is

$$h = 20(t) = 20(7.625)/12 = 12.7 \text{ ft}$$

We persist with the possibility of the 8-in. column, assuming the unsupported height to be 12.5 ft. We then proceed to use the *UBC* load formula, trying a column with four Number 7 bars.

$$P = A_g \left(0.18 f'_m + 0.65\, p_g f_s\right) \left[1 - \left(\frac{h}{40t}\right)^3\right]$$

$$= (7.625 \times 15.625) \left[(0.18)(1.5) + (0.65)\left(\frac{2.40}{119}\right)(20)\right]$$

$$\left[1 - \left(\frac{150}{40(7.625)}\right)^3\right]$$

$$= (119)(0.27 + 0.26)(0.88)$$

$$= 55.5 \text{ k}$$

Reducing this result by half because of the 8-in. dimension but increasing by one-third for wind, the actual design value is

$$P = 55.5 \times 0.5 \times 1.33 = 36.9 \text{ k}$$

which is considerably short of the design load of 87.9 k.

Assuming it to be acceptable architecturally, we widen the end of the wall by using a wider concrete block. Figure 6.61 shows such a detail. Assuming a 12-in. by 16-in. end column with four Number 7 bars, the new allowable load is

$$P = (11.625 \times 15.625) \left[(0.27) + 0.65\left(\frac{2.40}{182}\right)(20)\right.$$

$$\left.\left[1 - \left(\frac{150}{40(11.625)}\right)^3\right]\right]$$

$$= (182)(0.27 + 0.17)(0.97)$$

$$= 77.77 \text{ k}$$

which is more than adequate with the allowable stress increase.

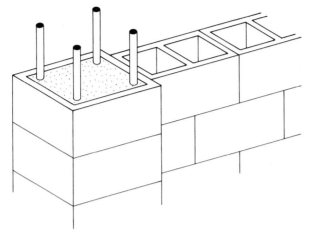

FIGURE 6.61 Optional shear wall end detail (Example 12).

Investigation of the first-story wall:

Shear stress: $\dfrac{66,000}{30} = 2200$ lb/ft

Overturning M: $13.5 \times 50 = 675$ k-ft

$+\ 26.25 \times 35 = 919$

$+\ 26.25 \times 20 = \underline{525}$

Total $= 2119$ k-ft

Dead load M: $170 \times 15 = 2550$ k-ft

Safety factor is less than 1.5.

Find:

Required resisting M: $1.5(2119) = 3179$ k-ft

Less dead load M: $\underline{2550}$

Required tie-down moment: 629 k-ft

Tie-down force at end of wall: $\dfrac{629}{28.67} = 21.9$ k

Compression force at end of wall: $\dfrac{2119}{28.67} = 73.9$ k (wind only)

If an 8-in. block is used, the shear stress on the net wall section is

$$v = \frac{2200}{12 \times 7.625 \times 0.45} = 53.5 \text{ psi}$$

With $M/Vd = 2119/(66)(28.67) = 1.12$, the maximum stress permitted by *UBC* Table 24–H is $1.33(35) = 47$ psi with no special inspection. With special inspection, the allowable stress is $1.33(1.5 \sqrt{1500}) = 77$ psi. The alternatives, therefore, are to have special inspection or to increase the thickness of the wall. This is probably a moot point, anyway, because the maximum allowable height of the wall is limited by *UBC* Table 24–I to 25 times the wall thickness, or $25(7.625)/12 = 15.88$ ft. With the 20-ft floor-to-floor height, it is likely that the unsupported height exceeds this, so a thicker wall is required.

With a 12-in. block the shear stress on the net wall section drops to

$$v = \frac{2200}{12 \times 11.625 \times 0.45} = 35 \text{ psi}$$

which is below the allowable stress without inspection.

For the end column design we assume the following conditions:

Gravity dead plus live load: 100 k

Total gravity plus wind load: $100 + 73.9 = 173.9$ k

Tension force for end anchorage: 21.9 k

Unsupported height: 18 ft

Use $f'_m = 1500$ psi, $f_y = 50$ ksi

The relatively low anchorage force is not critical because the dowels for the column reinforcing will easily provide this much anchorage. We therefore proceed to design the column for the compression force of 173.9 k. A 12-in. by 16-in. column requires considerable reinforcing, as an inspection of the previous calculation will show. Furthermore, the increased height results in additional reduction due to h/t. We therefore try for a 16-in.-square column with six Number 9 bars. Using the *UBC* formula, the allowable axial load is

$$P = (15.625)^2 \left[(0.18)(1.5) + 0.65\left(\frac{6.00}{244}\right)(20) \right] \left[1 - \left(\frac{18 \times 12}{40(15.625)}\right)^3 \right]$$

$$= 244(0.27 + 0.32)(0.96)$$

$$= 138 \text{ k}$$

With the allowable increase for wind the axial load is increased to

$$P = 1.33(138) = 184 \text{ k}$$

Consideration should be given to the resistance of the walls to horizontal sliding. Resistance is partially developed by a combination of friction and bonding of the mortar at the base of the walls, which may be considered reasonably adequate as long as the dead load is considerably higher than the horizontal load, which is the case in this example. However, some designers consider it desirable to provide some more positive type of anchorage.

One possibility is to use a series of shear keys, similar to those used for concrete walls in similar situations. If the masonry walls sit on concrete, the shear keys may be formed with wood blocks inserted in the top of the concrete during pouring and removed after the concrete has hardened. Figure 6.62 shows such a detail, with the keys located to correspond to the filled blocks in the masonry wall. Because there will also be reinforcing dowels at these points, the blocks may be drilled and sawn in two as shown in the illustration for accommodation of the dowels.

For a very conservative design the keys would be designed for end bearing and pure shear on the key throat to develop the entire horizontal wind force.

Another consideration for this example is that of the effect of the combined gravity and wind loads on the structure beneath the first-floor level. We consider two possibilities for this situation: with and without a basement. Figure 6.63 shows the situation in

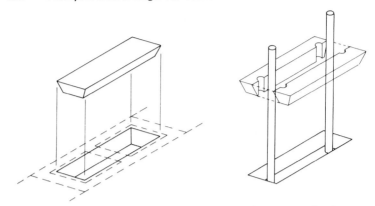

FIGURE 6.62   Forming the horizontal shear keys (Example 12).

which the shear wall is extended down another story as a basement wall and rests on a footing beneath the basement floor. In this case the resistance to the horizontal force of the wind is developed at the first floor, being transmitted through the first floor structure to the exterior basement walls and into horizontal soil pressure on the walls.

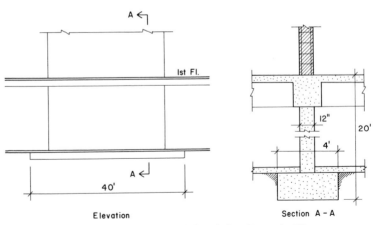

FIGURE 6.63   Shear wall foundation (Example 12).

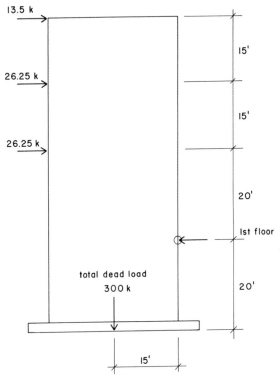

FIGURE 6.64 Overturn analysis of the shear wall (Example 12).

Figure 6.64 illustrates an overturn analysis of the shear wall foundation based on these assumptions. The pivot point for overturn is assumed to be at the first floor. This makes the overturning moment the same as that calculated previously for the first story shear wall: 2119 k/ft. However, the dead load resisting moment in this case includes the weight of the foundation. Thus

$$\text{Dead load } M: (300 \text{ k})(15 \text{ ft}) = 4500 \text{ k-ft}$$

and the safety factor against overturn is:

$$SF = \frac{4500}{2119} = 2.12$$

The soil stress in this case is limited to vertical pressure caused by gravity and horizontal sliding caused by the wind load. The vertical pressure is

$$p = \frac{300}{40 \times 4} = 1.875 \text{ ksf}$$

and the required sliding resistance force is

$$F = \frac{2119}{20} = 106 \text{ k}$$

which requires a soil friction coefficient of

$$\frac{106}{300} = 0.353$$

which is quite high, so it may be necessary to rely on ties to carry some of the required friction force to other foundations of the building.

The second condition we consider for the foundation is that which occurs when there is no basement. Figure 6.65 shows details for such a condition, in which the base of the wall consists

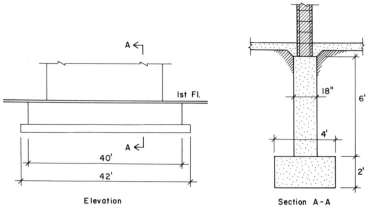

FIGURE 6.65   Shear wall foundation without basement (Example 12).

of a grade beam resting on a footing. In this case the first floor is ignored, and the wall is analyzed as a free-standing wall on its footing. The overturn analysis, as shown in Figure 6.66, is thus as follows:

Overturn M:

$$13.5 \times 58 = 783 \text{ k-ft}$$
$$+ \ 26.25 \times 43 = 1129$$
$$+ \ 26.25 \times 28 = \underline{\ 735}$$
$$\text{Total} \qquad = 2647 \text{ k-ft}$$

Dead load M: $270 \times 21 = 5670$ k-ft

Safety factor against overturn: $\dfrac{5670}{2647} = 2.14$

For soil pressure analysis, as shown in Figure 6.66, the over-turning moment of 2647 k–ft produces an equivalent eccentricity of

$$e = \frac{2647}{270} = 9.8 \text{ ft}$$

Because this is outside the kern of the section, we use the cracked section analysis for the maximum soil pressure, as shown in Figure 6.66. (See discussion in Section 6.20.)

$$\text{Pressure volume} = \frac{1}{2}(33.6)(4)(p) = 270 \text{ k}$$

$$\text{Thus: } p = \frac{2(270)}{(33.6)(4)} = 4.02 \text{ ksf}$$

Because the horizontal wind force is not resisted by the first floor in this case, it must be developed as friction on the bottom of the footing. The required value for the coefficient is thus:

$$\frac{66}{270} = 0.244$$

From *UBC* Table 29-B we see that obtaining such friction is possible with any soil other than one with a predominantly clay content.

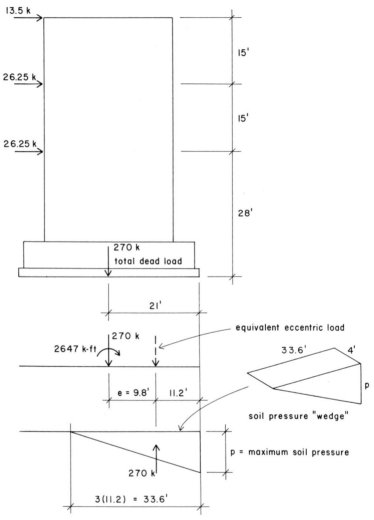

FIGURE 6.66 Analysis of the shear wall foundation (Example 12).

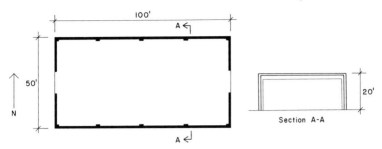

FIGURE 6.67 Building (Example 13).

6.13 Example 13

Figure 6.67 shows a simple one-story building in which the laterally resistive structural system that resists wind in the north–south direction consists of a series of single bent rigid frames that occur at each end of the building and at 25-ft intervals throughout the length. Assuming the walls to span vertically, the edge load to the roof diaphragm with an assumed wind pressure of 20 psf is

$$w = 1/2 \times 20 \times 20 \text{ psf} = 200 \text{ lb/ft}$$

If the roof diaphragm is flexible, this load is distributed to the bents on a peripheral basis. Thus, the bent loads are

End bent: $12.5 \times 200 = 2500$ lb/bent

Intermediate bent: $25 \times 200 = 5000$ lb/bent

If the roof diaphragm is considered to be reasonably rigid, it tends to distribute the loads to the bents in proportion to the bent stiffnesses. Thus, if they are of approximately equal stiffness, the load on each bent is

$$\frac{200 \times 100}{5} = 4000 \text{ lb/bent}$$

We assume the latter to be the case and design the bents for the 4000-lb load, applied as a concentrated load at the top of the bent.

Figure 6.68 shows the free body diagram of the bent with the wind loads and reactions. The deflected shape under wind load is shown by the dashed line. Although this problem is essentially

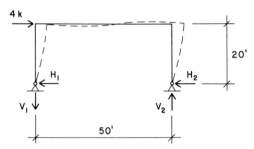

FIGURE 6.68 Function of the bent for wind load (Example 13).

indeterminate, if we assume the bent to be symmetrical, we may reasonably assume the two horizontal reactions to be equal. In any event the vertical reactions are statically determinate and may be determined as follows:

$$V_1 = V_2 = \frac{4 \times 20}{50} = 1.6 \text{ k}$$

On the basis of this analysis for the reactions, the distribution of internal forces is shown in the illustrations in Figure 6.69.

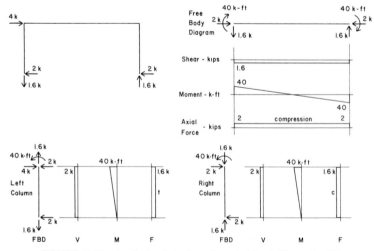

FIGURE 6.69 Analysis of the bent for wind load (Example 13).

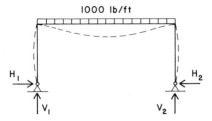

FIGURE 6.70 Function of the bent for gravity load (Example 13).

These forces must be combined with the forces caused by the gravity load to determine the critical design conditions for the bents.

Figure 6.70 shows the bent as loaded by a uniform load of 1000 lb/ft on top of the horizontal member. This effect is based on an assumption that the roof framing delivers an approximately uniform loading with a total dead plus live load of 40 psf as an average for the roof construction, including the weight of the horizontal bent member. As with the wind load, the vertical reactions may be found on the basis of the bent symmetry to be:

$$V_1 = V_2 = \frac{1 \times 50}{2} = 25 \text{ k each}$$

Determination of the horizontal reactions in this case, however, is indeterminate and must consider the relative stiffness (I/L) of the bent members. For the pin-based columns it will be found that the horizontal reactions will each be:

$$H = \frac{w\,L^3\,I_c}{8\,I_g h^2 + 12\,I_c\,hL}$$

In this calculation we may use the relative, rather than actual, values of the column and girder stiffness (I_c and I_g in the formula). If we assume the girder to be approximately 1.5 times as stiff as the column, the horizontal reaction will be:

$$H = \frac{(1)(50)^3(1)}{8(1.5)(20)^2 + 12(1)(20)(50)} = \frac{125,000}{4800 + 12,000} = 7.44 \text{ k}$$

With these values for the reactions, the free body diagrams and distribution of internal forces for the gravity loading are as shown in Figure 6.71. These must next be combined with the previously determined wind forces, as shown in Figure 6.72. The design conditions for the individual bent members would be selected from the maximum values of Figure 6.71 (gravity only) or three-quarters of the maximum values of the combined loading, as shown in Figure 6.72. The adjustment of three-quarters for the comparison is based on the increased allowable stress for the combined forces that include the wind load.

We may now proceed to make a preliminary design of the column and girder, based on these analyses. For the illustration we will design the bent in steel, using standard rolled sections, and then discuss other possibilities for its construction. For the column the critical condition is that of the leeward column, which must be designed for the compression plus bending. We assume the following for the first trial design:

A36 steel, rolled WF sections

Design for $M = 148.8$ k-ft, axial compression $= 25$ k

(Combined loading is not critical; no increase in allowable stresses.)

Assume the wall to brace the column continuously on its y axis.

We may now try a section by picking one from the S_x tables in the American Institute of Steel Construction Manual (Ref. 5) with a moment capacity slightly higher than 148.8 k-ft, based on the assumption that the axial load is not very important because of the relatively high stiffness on the unbraced x axis. We thus will try a W 16 × 58, with a listed moment capacity of 189 k-ft with $F_y = 36$ ksi. Checking this with Formula 1.6–2 from the steel design specification of the AISC manual, we get:

$$\frac{f_a}{F_a} + \frac{f_b}{F_b} + \leq 1$$

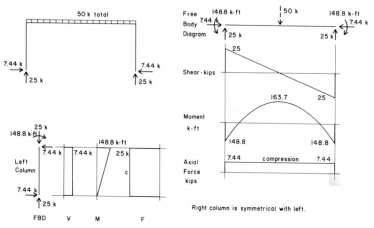

FIGURE 6.71 Analysis of the bent for gravity load (Example 13).

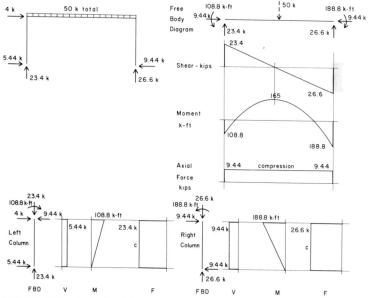

FIGURE 6.72 Analysis of the bent for combined wind and gravity loads (Example 13).

Where: $f_a = \dfrac{25}{17.1} = 1.46$ ksi

$$\frac{KL}{r_x} = \frac{2(240)}{6.62} = 72.5$$

$F_a = 16.17$ ksi (from AISC Manual, p. 5–84)

$$f_b = \frac{M}{S} = \frac{(148.8 \times 12)}{94.4} = 18.92 \text{ ksi}$$

$F_b = 24$ ksi

Then

$$\frac{f_a}{F_a} + \frac{f_b}{F_b} = \frac{1.46}{16.17} + \frac{18.92}{24} = 0.09 + 0.79 = 0.88$$

which indicates that the section is adequate on a stress basis.

There are two additional design considerations for the column that have some importance. One has to do with the connection of the column to the girder. If the connection is to be a fully welded connection, the flange widths of the two members should be reasonably matched. The other consideration has to do with the lateral deflection, or drift, of the bent, which is discussed later.

For the girder design we assume the following:

A36 steel, rolled WF section.
Design for M: 163.7 k-ft (positive moment at midspan).
Axial compression is negligible, based on the column design.
Assume roof framing braces the girder on 6-ft centers.

Once again using the S_x tables from the AISC manual, we try a W 21 × 49, with a listed moment capacity of 187 k-ft. This will make the f_b/F_b ratio in the combined stress formula:

$$\frac{163.7}{187} = 0.875$$

Then: $f_a = \dfrac{7.44}{14.4} = 0.517$ ksi

$$\frac{KL}{r_y} = \frac{1(72)}{1.31} = 55$$

$$F_a = 17.90 \text{ ksi}$$

which clearly indicates that the combined stress is not critical.

Two deflection problems should be considered. The first is that of the vertical deflection of the girder, which is most critical for the end bent because the details of the end wall construction must tolerate at least the deflection due to the live load. If we assume the live load to be approximately half of the total load, this deflection will be something less than

$$\frac{5\,WL^3}{384\,EI} = \frac{5(25)(50)^3(1728)}{384(30,000)(971)} = 2.4 \text{ in.}$$

which is the simple beam deflection, based on no end moments.

The end moment will produce an upward deflection equal to $ML^2/8\,EI$. Using half of the calculated end moment for the live load deflection, this reduction amounts to

$$\frac{ML^2}{8\,EI} = \frac{(74.4)(50)^2(1728)}{8(30,000)(971)} = 1.38 \text{ in.}$$

resulting in a net midspan deflection of approximately 1 in.

The other deflection consideration is that of the lateral drift caused by wind, as mentioned previously. Again, this is most critical for the end wall construction. As shown in Figure 6.73, the lateral deflection (Δ in the figure) may be calculated in two parts. The first part consists of the simple cantilever deflection of the column (t_1 in the figure). The second part is caused by the rotation at the top of the column (t_2 in the figure). The determination of this combined deflection is

$$\Delta = t_1 + t_2$$

$$= \frac{Hh^3}{3\,EI_c} + \frac{Hh^2\,L}{8\,EI_g}$$

$$= \frac{(2)(20)^3(1728)}{3(30,000)(748)} + \frac{(2)(20)^2(50)(1728)}{8(30,000)(971)}$$

$$= 0.41 + 0.30 = 0.71 \text{ in.}$$

Although these are live load deflections and consequently quite theoretical, the end wall construction, as well as any interior cross wall construction, should be designed to tolerate movements of this order of magnitude.

Connection of the column and girder rolled sections are done with some combination of bolting and welding. Figure 6.74 shows two possibilities, both based on the assumption that the bent cannot be transported from the fabricating shop to the site in one piece. In the upper part of the figure, welded plates extend beneath the girder and are bolted to each side of the column. In the lower part of the figure, the column-to-girder connection is fully welded, and the field connection consists of a splice in the girder at the approximate location of the inflection point under gravity load. The web plates of this connection are designed for the girder

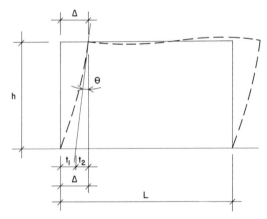

FIGURE 6.73   Bent deflection caused by wind load (Example 13).

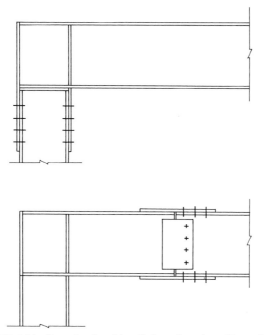

FIGURE 6.74 Bent details with rolled steel sections (Example 13).

shear and the flange plates are designed for the wind moment plus
some amount of the gravity moment because unbalanced live load
would produce some moment at this theoretical zero moment
point.

Although we do not illustrate their design, Figure 6.75 shows
details for four other possible constructions of the bents. Figure
6.75(a) shows a bent built up from flat plates of steel, producing
the same basic I-shaped cross section for the members. An advantage of this construction, as well as of the others in the figure,
is that the members may be tapered in length. This offers the
possibility of designing the girder for the required depths at the
midpoint and ends and the possibility of providing for roof drainage while maintaining a flat bottom on the girder. And it offers the
advantage of reducing the size of the column at the floor level
where the moment capacity is not required.

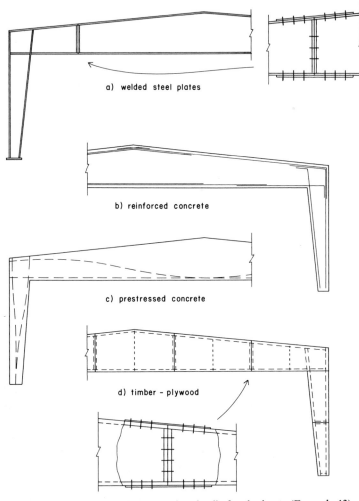

a) welded steel plates

b) reinforced concrete

c) prestressed concrete

d) timber - plywood

FIGURE 6.75 Optional construction details for the bents (Example 13).

Figures 6.75(b) and 6.75(c) illustrate construction of the bent in concrete with either conventional reinforcing or prestressing with steel cables. If the bents are cast flat on the site and tilted in place, they could be built in one piece.

Figure 6.75(d) shows the possibility of a plywood and timber box-type construction for the bent. Such bents would most likely be primarily shop-built, using field splices similar to those for the steel bents. The sketch in the figure shows a possibility for a joint placed near the girder inflection point.

6.14   Example 14

This example consists of the same building plan and section as used in Example 2. The modification in this case consists of developing the front wall as a multibay rigid frame. The east–west wind load is therefore shared by the rear shear walls and the rigid frame, as shown in Figure 6.76.

There are two approaches to the determination of the distribution of the load to the two walls. The first is to assume a peripheral distribution with half of the total load carried by each wall. By referring to the calculations in Example 2, one can see that

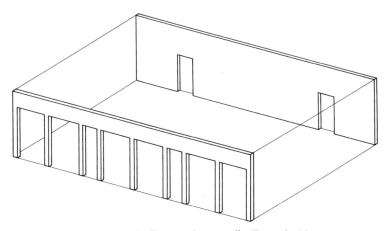

FIGURE 6.76   Front and rear walls (Example 14).

this results in a load of 2400 lb for each wall. The second approach is to determine the relative stiffnesses of the two walls, to locate the center of stiffness, and to do a combined direct force and torsion analysis. This approach requires that the actual deflections of the two walls be determined to compare their stiffnesses.

Because the deflection of the rigid frame cannot be determined unless the sizes of the members are known, the simpler peripheral distribution would probably be used as a basis for approximate sizing of the frame, even if the second analysis is performed. We illustrate the design with the peripheral distribution, and then discuss the second approach.

Both the horizontal and vertical members of the rigid frame must be designed for the combined gravity and wind loadings. The columns are designed for combined axial compression plus bending. The beams are essentially designed for the combined wind and gravity moments because the roof diaphragm probably provides sufficient lateral bracing to make the axial compression negligible.

The approximate wind shear in the columns may be found by assuming them to act as vertical cantilevers, as shown in Figure 6.77. We assume the columns to be all one size. The end columns

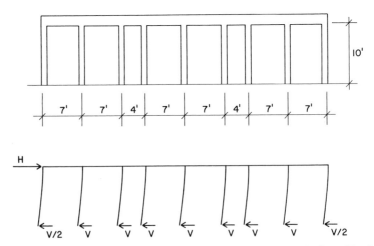

FIGURE 6.77   The front wall bent and its assumed behavior under lateral load (Example 14).

are somewhat less stiff, however, because there is less resistance to rotation at their tops because of the discontinuity of the beam. For the approximate analysis, we assume the interior columns to be twice as stiff as the end columns. The seven interior columns plus the two end columns thus add up to eight equivalent interior columns, and the shear on an interior column is

$$V = \frac{2400}{8} = 300 \text{ lb}$$

and the wind moment on the cantilevered column is

$$M = 300(10 \text{ ft}) = 3000 \text{ lb-ft}$$

If we try a W 6 × 20 with an S = 13.4 in^3 and A36 steel with F_y = 36 ksi, then

$$f_b = \frac{M}{S} = \frac{3000(12)}{13.4} = 2687 \text{ psi}$$

$$\frac{f_b}{F_b} = \frac{2687}{1.33(24,000)} = 0.084$$

which leaves a considerable margin for the axial compression.

Once again, assuming the columns to function as vertical cantilevers, the approximate lateral deflection of the frame can be found by using the derived deflection formula for a cantilever with a concentrated load at the end.

$$\Delta = \frac{PL^3}{3 \, EI} = \frac{(300)(10 \times 12)^3}{(3)(29.6 \times 10^6)(41.5)} = 0.14 \text{ in.}$$

The actual deflection is somewhat more than this because of the rotation of the top of the column caused by the flexing of the beam. To be safe the detailing of the front wall should be capable of tolerating a deflection of 0.25 in. or so. Otherwise, the column size should be increased, even though the frame will be oversized on a stress basis.

Figure 6.78 shows a scheme for the fabrication of the frame as a series of three prefabricated two-span bents with a short shear connected member at the door opening. Figure 6.79 shows the details for such a frame with the columns welded to the bottom of the beams and web plates added to the beam to carry the frame moments to the top flange of the beam.

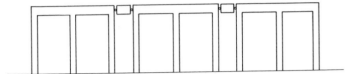

FIGURE 6.78   Optional forming of the front wall bent (Example 14).

With the frame members approximately sized on the basis of the simple peripheral distribution, a more rigorous analysis for the lateral loads can now be performed if it is deemed necessary by the designer. The first step in this analysis would be to find the deflection of the two walls caused by a unit load. Comparison of the two deflections would provide a basis for establishing the relative stiffnesses of the walls. With this determined, the analysis would then proceed as in the analysis of the masonry walls in Example 3.

6.15   Example 15

Figure 6.80 shows a plan and section for a building similar in form to that in Example 12. In this case, however, the laterally resistive system for north–south wind consists of seven three-story, three-bay rigid bents. The exact analysis of this type of structure is beyond the scope of this work. We illustrate an approximate

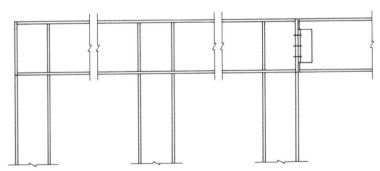

FIGURE 6.79   Framing details for the front wall bent (Example 14).

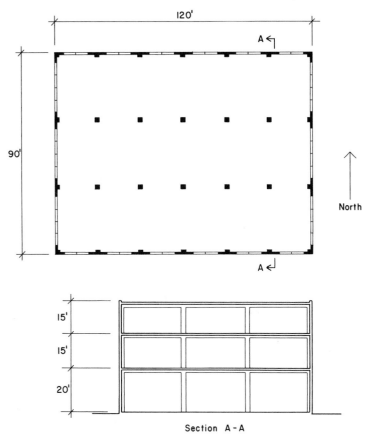

Section A - A

FIGURE 6.80 Building (Example 15).

analysis that may be used for preliminary design and for rough cost estimates.

Assuming the roof and floor diaphragms to be reasonably stiff, and the bents to be approximately equal in stiffness, the total wind load may be divided equally between the seven bents. On this basis, the bent loading will be as shown in Figure 6.81. Assuming the inflection points of both columns and beams to be at the midpoint of the spans, except for the first-story columns, the

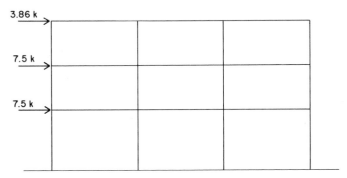

FIGURE 6.81   North–south wind load on the three-story bent (Example 15).

internal shear forces are as shown in Figure 6.82. The end moments in the members may then be determined from these shears.

These approximate wind moments must be combined with the moments and axial loads caused by gravity for the design of the beams and columns in the bents. These preliminary sizes may then be used for a more exact indeterminate analysis. An excellent discussion of the various problems of analysis and design of multistory framed structures may be found in *High-Rise Building Structures* (Ref. 15).

6.16   Example 16

The plan and framing elevation shown in Figure 6.83 show a one-story building in which the lateral resistive system consists of a braced frame. For east–west wind load, the frame is braced by the X-braces in the end bays of the front and rear walls. With a design wind pressure of 25 psf the total wind load on the end of the building is

$$(25 \text{ psf})(20 \times 50) = 25,000 \text{ lb}$$

Assuming the end wall to span vertically, half of the total wind load is delivered to the edge of the roof. Assuming a distribution to the front and rear walls on a peripheral basis, the total load on each wall ("H" in the figure) is 6250 lb. With two X-braced bays

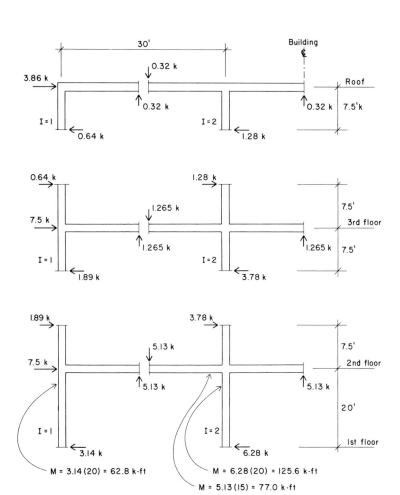

FIGURE 6.82 Approximate analysis of the bent for north–south wind load (Example 15).

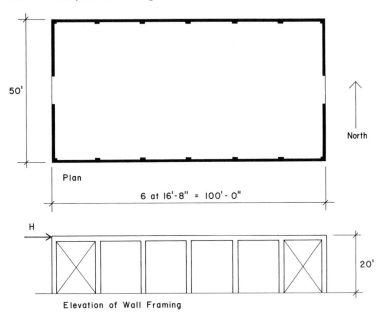

FIGURE 6.83   Building (Example 16).

in each wall, the loading of each braced bay is as shown in Figure 6.84.

Although the X-brace essentially produces an indeterminate structure, the usual practice for static load is to assume that the relative slenderness of the diagonals will result in the buckling of the compression member at a very low load, after which the total load is assumed to be taken by the tension member. For the reversible wind loading, both members are thus designed as tension members, with each working for wind in a single direction.

With the geometry of the brace as shown in Figure 6.84, the tension force in the brace is

$$T = 1.56(3125) = 4875 \text{ lb}$$

Using a round steel rod of A36 steel, this force could be developed by a 5/8 in. diameter rod on the basis of stress on the net section of the threaded rod and the allowable one-third increase in

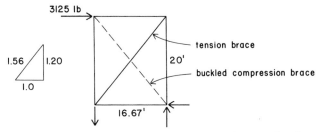

FIGURE 6.84   Function of the X-braced frame (Example 16).

stress from wind. For reasons of the wall construction details or the development of the connections of the braces to the frame, it may be desirable to use something other than a round rod, however. Also, to reduce the total wind deflection of the frame, it may be desirable to oversize the braces to reduce the magnitude of their tension elongation.

6.17   Example 17

Figure 6.85 shows the roof framing plan for a large one-story building in which the horizontal diaphragm consists of a braced frame. With the scheme shown, the wind load in each direction is

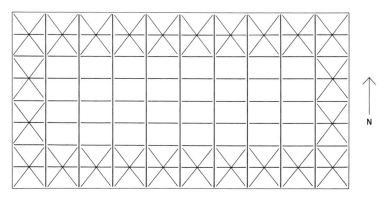

FIGURE 6.85   Roof framing plan (Example 17).

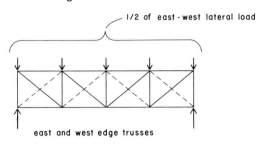

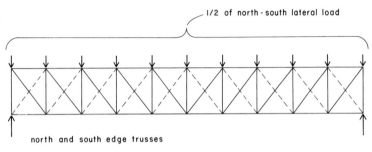

FIGURE 6.86 Functions of the roof edge trusses (Example 17).

resisted by two trusses, one on each side of the building. For wind in a single direction, only half of the diagonals (the tension members) are assumed to function. Thus, the two types of trusses are as shown with their loads in Figure 6.86.

The members of the gravity load resistive system (as shown on the rectangular grid in the plan) function as chords and as interior compression members for the trusses. On the interior of the roof, between the trusses, the gravity resistive members act as drag struts to carry the wind across the roof to the opposite trusses. Whereas the diagonal braces function for wind alone, the other frame members must be designed for the combined wind and gravity loads.

It is theoretically possible to reduce the number of diagonal braces by using the framing scheme as shown in Figure 6.87. In this case only one truss works for the wind in a single direction. Thus, the force per brace, and the combined forces in the gravity

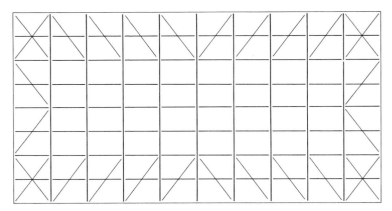

FIGURE 6.87   Optional bracing of the roof frame (Example 17).

resistive members, are higher. In most cases designers prefer the X-braced system for this type of structure.

6.18   Example 18

This example consists of the same building as shown in Example 12. In this case the three-story-high shear walls are replaced by X-braced bays, as shown in Figure 6.88. The wind loading for these braced frames and the geometry of the braces are shown in Figure 6.89. For the calculations of wind load, see Example 12.

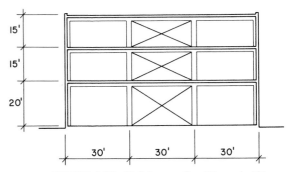

FIGURE 6.88   Building section (Example 18).

Assuming the braces to be of A36 steel, the forces and required areas for the braces are as follows:

Third story:

Story shear: 27 k (total for the two bays)

Brace tension: $\dfrac{2.236}{2}(27)\left(\dfrac{1}{2}\right) = 15.1$ k

Required area: $\dfrac{15.1}{1.33(21.6)} = 0.526$ in²

Second story:

Story shear: 79.5 k

Brace tension: $\dfrac{2.236}{2}(79.5)\left(\dfrac{1}{2}\right) = 44.4$ k

Required area: $\dfrac{44.4}{1.33(21.6)} = 1.55$ in²

First story:

Story shear: 132 k

Brace tension: $\dfrac{3.606}{3}(132)\left(\dfrac{1}{2}\right) = 79.3$ k

Required area: $\dfrac{79.3}{1.33(21.6)} = 2.76$ in.²

The type of member to be used for the braces depends on the problems of integrating the brace into the wall construction and connecting it to the steel frame. When the area requirements are higher than can be developed with small diameter rods, a common choice is that of single angles. If the angle legs are turned in opposite directions in the bay, they can be lapped and fastened at their intersection for simple mutual bracing.

As discussed for the previous examples of X-braced frames, it may be desirable to oversize the braces in order to reduce the frame deflection.

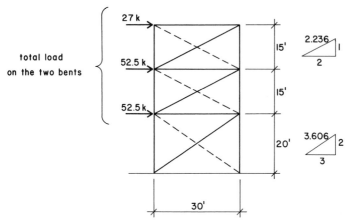

FIGURE 6.89   Loading of the three-story braced frame (Example 18).

6.19   Example 19

Figure 6.90 shows a section through a reinforced concrete block wall that is free-standing and that is designed for a wind pressure of 20 lb/ft² and the following criteria:

Maximum soil pressure: 1500 lb/ft²

Maximum tension in reinforcing: 20,000 psi

f'_m = 1500 psi for the concrete block

With no special inspection, *UBC* Table 24–H gives the following for the reinforced masonry:

$$n = 40, f_m = 250 \text{ psi}$$

The design of the wall consists of determining the thickness of the block, the size and spacing of the vertical reinforcing, and the dimensions and reinforcing (if any) for the footing. In many situations, these factors can all be established from predesigned tabulations in various handbooks and detailing manuals. With no such reference available, the design is essentially a matter of trial and error. For a first try we assume an 8-in. block with reinforcing placed in concrete-filled voids at 48-in. centers, which gives an

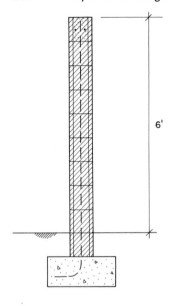

FIGURE 6.90 Section (Example 19).

average weight for the wall with sand and gravel block of 60 lb/ft^2 of the wall surface.

With a 10-in. thick footing placed with its base 18 in. below grade, the loading analysis is as shown in Figure 6.91. At the base of the wall (top of the footing) the moment caused by wind is

$$M = (120)(3.67) \text{ ft} = 440 \text{ lb-ft}$$

This moment is used to determine the maximum compressive stress in the masonry and to find the required reinforcing. Assuming the reinforcing bars to be centered in the wall, the effective depth of the cross section is half the wall thickness, or approximately 3.8 in. Because an exact analysis of the stresses in the reinforced section cannot be made until the area of the reinforcing is known, we first use an estimated value for j to find a trial steel area.

If $j = 0.9$

$$A_s = \frac{M}{f_s\,jd} = \frac{(440)(12)}{(1.33)(20{,}000)(0.9)(3.8)} = 0.058 \text{ in}^2/\text{ft}$$

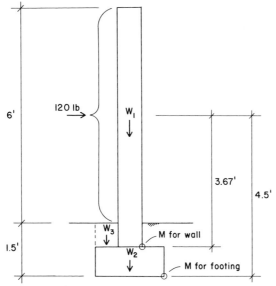

FIGURE 6.91 Wind and gravity loading (Example 19).

Thus, at 48-in. centers the bar area must be:

$$A_s = (4)(0.058) = 0.232 \text{ in}^2$$

which requires that it be a Number 5 bar with $A_s = 0.31$ in².

Using this steel area it is now possible to determine the percentage of steel in the cross section and to use this with the value of n to find the k and j for the wall. These may be found by using formulas or handbook tables or graphs, such as those in the *Concrete Masonry Design Manual* (Ref. 7.) From these data it will be found that

$$k = 0.307, \quad \text{and} \quad j = 0.898$$

It is not necessary to check the stress in the reinforcing because the bar selection provides considerably more than the calculated area required. We proceed, therefore, to check the maximum compressive stress in the masonry, as follows:

$$f_m = \frac{M}{bd^2}\left(\frac{2}{kj}\right) = \frac{(440)(12)(2)}{(12)(3.8)^2(0.307)(0.898)} = 221 \text{ psi}$$

From *UBC* Table 24–H the allowable flexural stress in compression, without special inspection, is

$$f_m = (0.166)(f_m) = (0.166)(1500) = 249 \text{ psi}$$

Thus, the stress is not critical, even without considering the one-third increase permitted for wind.

Referring to Figure 6.91, if the footing is 2 ft wide and 10 in. thick, the overturn analysis is

$$W_1 = (60 \text{ lb/ft}^2)(6.67 \text{ ft}) = 400 \text{ lb}$$

$$W_2 = (150 \text{ lb/ft}^3)(2)\left(\frac{10}{12}\right) = 250 \text{ lb}$$

$$W_3 = (100 \text{ lb/ft}^3)\left(\frac{8}{12}\right)\left(\frac{8}{12}\right) = 44 \text{ lb}$$

Overturn M: $(120)(4.5) = 540$ lb-ft
Dead load M: $(400)(1)\quad = 400$
$\qquad\qquad\quad (250)(1)\quad = 250$
$\qquad\qquad\quad (44)(1.67) = \underline{73}$
$\qquad\qquad$ Total $\quad\ \ = 723$ lb-ft

Safety factor against overturning: $\dfrac{723}{540} = 1.34$

Because this is close to but not quite equal to the usual requirement for a safety factor of 1.5, we try a second time with a footing width of 27 in., which will slightly increase the footing dead load and the moment arms for all three dead loads.

New W_2: $(150)\left(\dfrac{27}{12}\right)\left(\dfrac{10}{12}\right) = 281$ lb

New dead load M: $(400)(1.125) = 450$ lb-ft
$\qquad\qquad\qquad\ \ (281)(1.125) = 316$
$\qquad\qquad\qquad\ \ (53)(1.85)\quad = \underline{98}$
$\qquad\qquad\qquad$ Total $\qquad\ = 864$ lb-ft

New safety factor: $\dfrac{864}{540} = 1.60$

The remaining determination to be made is whether the combination of vertical load and wind moment stresses exceeds the allowable soil pressure. With the wind included, the permitted maximum soil pressure is 1.33(1500) = 2000 psf. The vertical load to be used is the sum of the weights of the wall and footing: 681 lb. With the 27-in.-wide footing, this load produces a stress of

$$p = \frac{681}{2.25} = 303 \text{ psf}$$

The wind moment of 540 lb-ft results in an equivalent eccentricity of

$$e = \frac{M}{N} = \frac{(540)(12)}{681} = 9.52 \text{ in.}$$

which is considerably outside the kern of the section (1/6 the width, or 4.5 in.). Therefore, a so-called *cracked section analysis* must be done. The basis for this analysis is illustrated in Figure 6.92. The wedge-shaped pressure volume represents the total force exerted by the soil. This force must be equal to N (681 lb) and must be colinear with the equivalent eccentric N, as shown, for static equilibrium. Because e has been determined, the value of x in the figure is known, and the volume can be expressed in terms of the three dimensions of the wedge. The resultant force V is located at the centroid of the volume, giving the third dimension of the wedge of $3x$. Thus

$$x = \frac{27}{2} - e = 13.5 - 9.52 = 3.98 \text{ in.}$$

$$V = N = 681 \text{ lb}$$

$$681 = \text{volume of the wedge}$$

$$= \frac{1}{2}(12)(3)(3.98)(p)$$

$$p = 9.51 \text{ psi}$$

$$= (9.51)(144) = 1369 \text{ psf}$$

which is less than the maximum permitted stress.

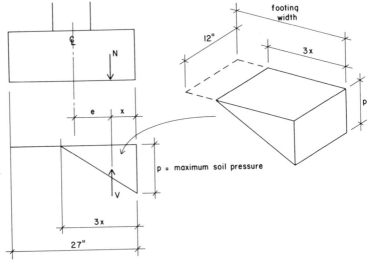

FIGURE 6.92 Analysis of footing (Example 19).

Notice that for fences *UBC* Section 2311(h) permits a design wind pressure of two-thirds of that given in Table 23-F of the *UBC*.

6.20 Example 20

As shown in Figure 6.93, this example consists of a tall, round tower, such as a smokestack, that is supported by a square, shallow, bearing-type footing. For the example we consider only the problems of overturning and of the maximum soil pressure caused by wind plus gravity loadings. We assume the following:

> Dead load of the tower: 60 k
> Wind pressure: 20 psf (30-ft reference height)

The wind and gravity loads are as shown in Figure 6.94. The weight of the footing is determined as

$$W = (150 \text{ lb/ft}^3)(12)(12)(2.5) = 54,000 \text{ lb or } 54 \text{ k}$$

From the *UBC* Table 23–G we derive the multiplying factor for

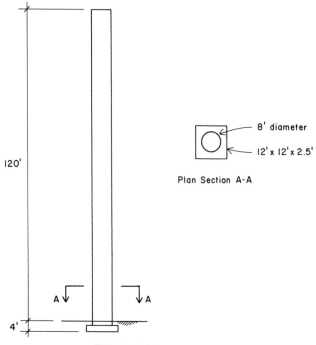

8' diameter

12' x 12' x 2.5'

Plan Section A-A

120'

A

A

4'

FIGURE 6.93 Example 20.

the round tower—0.60. Thus, the horizontal wind forces for the various pressure zones are calculated as

$$H_1 = (0.60)(30 \text{ psf})(8 \text{ ft})(20 \text{ ft}) = 2880 \text{ lb}$$

$$H_2 = (0.60)(25)(8)(50) = 6000 \text{ lb}$$

etc.

The calculations for the wind loads and overturning moment are shown in Table 6.4. The total dead load resisting moment is

$$M = (114 \, k)(6 \text{ ft}) = 684 \text{ k-ft}$$

Because this is less than the overturning moment, the foundation is obviously not adequate.

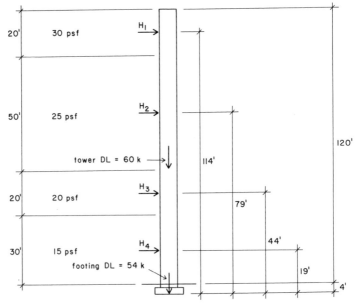

FIGURE 6.94 Wind loads (Example 20).

TABLE 6.4 Wind Moment on Footing: Example 20

Wind pressure zone	Total force (kips)	Distance from base (feet)	Moment (kip-ft)
1	$H_1 = 2.88$	114	328.3
2	$H_2 = 6.0$	79	474.0
3	$H_3 = 1.92$	44	84.5
4	$H_4 = 2.16$	19	41.0
Total shear	$= 12.96$ k	Total Moment $=$	927.8 k-ft

216

If the footing plan size is increased to 16 ft square, the new footing weight will be

$$W = (150)(16)^2(2.5) = 96{,}000 \text{ lb, or } 96 \text{ k}$$

and the new dead load resisting moment will be

$$M = (156)(8) = 1248 \text{ k-ft}$$

which results in a new safety factor of

$$SF = \frac{1248}{927.8} = 1.35$$

This result is still lower than the desired factor of 1.5. However, we have ignored the dead load of the soil over the footing in computing the dead load resisting moment. At an assumed density of 100 lb/ft³, this squared doughnut-shaped volume has a weight of

$$W = (100)\,[(16)^2(1.5) - \pi(4)^2(1.5)] = 30{,}860 \text{ lb}$$

Adding this load to the dead loads of the tower and footing, the total dead load becomes approximately 187 k, and the dead load resisting moment increases to

$$M = (187)(8) = 1496 \text{ k-ft}$$

which results in a safety factor of

$$SF = \frac{1496}{927.8} = 1.61$$

Whether to include the soil weight in this stability analysis or not is a matter of judgment. The other option is to increase the plan size and/or the thickness of the footing until the desired dead load moment is achieved.

With the total tower and footing weight of 156 k, the equivalent eccentricity with the wind moment is:

$$e = \frac{M}{N} = \frac{(927.8)}{156} = 5.95 \text{ ft}$$

Referring to the pressure wedge analysis illustrated in Example 19, this calculation produces a value for x of $8 - 5.95 = 2.05$ ft.

Thus, the maximum soil pressure is determined to be

$$N = 156 = V = 1/2(16)(3)(2.05)(p)$$
$$p = 3.17 \text{ k/ft}^2$$

6.21 Example 21

This example is similar to Example 20, except that the tower consists of an open truss frame, as shown in Figure 6.95. From the *UBC* Table 23–H we get the modifying factor for the open frame of 2.20. This factor is to be multiplied by the actual profile of the solid members of the frame. If we assume the member profiles to be 15% of the overall tower profile, the factor for wind load becomes 2.20(0.15), or 0.33. With the same tower profile and foundation as shown for the round tower in Figure 6.93, the overturn for the open tower will be 0.33/0.60 or 55% of that

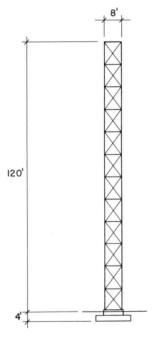

FIGURE 6.95 Example 21.

determined in Table 6.4 for the round tower. Thus

$$\text{Overturning } M: \frac{0.33}{0.60} (927.8) = 510 \text{ k-ft}$$

We will assume the weight of the open tower to be half of that for the round tower. If the 12-ft-square footing is used, the weight of the soil over the footing will be

$$W = (100)[(12)^2(1.5) - \pi(4)^2(1.5)] = 14{,}060 \text{ lb}$$

Thus, the total dead load of tower, footing, and soil is approximately 98 k, which results in a dead load resisting moment of

$$M = (98)(6) = 588 \text{ k-ft}$$

yielding a safety factor of only

$$SF = \frac{588}{510} = 1.15$$

This result means that the footing size must be increased. If the 16-ft-square footing from Example 20 is used, the total dead weight is increased to 157 k, and the dead load resisting moment becomes

$$M = (157)(8) = 1256 \text{ k-ft}$$

yielding a new safety factor of

$$SF = \frac{1256}{510} = 2.46$$

which is quite conservative for the overturn consideration alone. So if the soil pressure is sufficient and the design of the footing for concrete and reinforcing stresses permits it, the footing size could be less than 16 ft square.

Notice that the *UBC* Table 23–H also provides a load factor for the wind pressure applied to the diagonal profile of the square tower. In the table this is called the *load on the corner*. Adjustment factors are also given for towers with cylindrical members rather than flat-sided members.

7

Analysis and Design for Earthquake Forces

||||||||||||||||||||||||||||||||||||||||||||||||||||

Once the lateral forces have been determined, the design of the lateral resistive structural system is often essentially the same for both wind and earthquake effects. For this reason, and to shorten the work, many of the examples in this chapter consist of investigations of the buildings in examples in Chapter 6. Once the critical loads are determined, the design of typical horizontal diaphragms, shear walls, tie downs, and so on is procedurally similar for both load conditions. Although the magnitudes of the seismic loads occasionally require changes in the design of the systems or parts, in general we refer the reader to the examples in Chapter 6 for illustration of the basic design processes.

There are, of course, various special requirements for seismic design. Some of these are pointed out in the examples for comparison with the requirements for wind design. All buildings must sustain the critical load conditions required by their location. When both wind and seismic loads are high, it is not uncommon for some parts of the structure to be designed for wind and others for seismic forces.

7.1 Example 1

This example is the same as Example 1 in Chapter 6. The building plan and section can be seen in Figure 6.1. We assume the following for the seismic analysis:

Seismic zone 4, $Z = 1.0$. See Section 2312(c) of the *UBC*.

Occupancy importance factor: $I = 1.0$. See Table 23-K of the *UBC*.

S not determined, use maximum required CS of 0.14. See Section 2312(d) of the *UBC*.

Dead loads:

Roof and ceiling: 12 psf

Exterior walls: 20 psf (solid portions)

Interior walls: 10 psf

Canopy: 100 lb/ft

Rooftop HVAC units: 5000 lb total

For the box system building, *UBC* Table 23-I requires a K factor of 1.33. The total lateral seismic load on the building is therefore

$$V = Z I K C S W$$
$$= (1.0)(1.0)(1.33)(0.14)W$$
$$= 0.1862W$$

The critical loading is that which is applied to the edges of the roof diaphragm and is transferred through the shear walls to the foundations. For this load the value of W is the sum of the dead loads of the roof and ceiling, any items sitting on or suspended from the roof (such as HVAC units and light fixtures), the upper portions of the walls (assuming them to span vertically), and any items that are attached to the upper part of the walls (such as the canopy in this example.)

If a wall is of braced construction, even though it is not used as part of the building's lateral resistive system, it is usually assumed that it transfers load to the roof only in the direction that is perpendicular to the plane of the wall. This relationship was discussed in Chapter 2 and is illustrated in Figure 2.3. In this example, each of the four exterior walls is partially braced in its own plane. They are therefore assumed to add to the W summation only for the lateral load that is perpendicular to the wall plane.

The actions of the walls as assumed in this example are shown in Figure 7.1. In the direction perpendicular to their planes, the walls are assumed to span from floor to roof. For the interior walls, the load to the roof is half of the total dead load of the wall. For the exterior walls, we assume the location of the roof diaphragm to be at approximately 12 ft above the floor, so that the extension of the wall parapet is 2 ft above the roof. Assuming the function of the wall to be as described for Case 2 in Figure 6.18, we thus add a total of $12/2 + 2 = 8$ ft of wall dead load to the roof load.

We assume the canopy load to be added to W in both directions. We assume the HVAC units to be reasonably centrally located, so we will include their load in the W for both directions.

With these various assumptions, the tabulation of the total W for the lateral load in each direction is as shown in Table 7.1. Notice that we have assumed that all the exterior walls are solid, even though there are several door and window openings. We make this assumption because we are including the upper por-

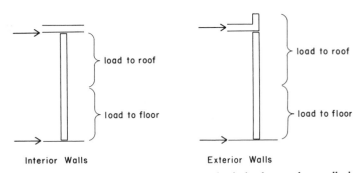

Interior Walls   Exterior Walls

FIGURE 7.1   Assumed wall functions with seismic load normal to wall plane (Example 1).

TABLE 7.1 Dead Load for the Roof Diaphragm Loading: Example 1

Source of Load	Unit Load	East–West Load		North–South Load	
Roof and ceiling	12 psf	$30 \times 50 \times 12 =$	18 k	$30 \times 50 \times 12 =$	18 k
Exterior Walls	20 psf	$60 \times 8 \times 20 \ =$	9.6	$100 \times 8 \times 20 =$	16
Partitions	10 psf	$100 \times 6 \times 10 =$	6	$100 \times 6 \times 10 =$	6
Canopy	100 lb/ft	50×100	5	50×100	5
HVAC units	5 k total		5		5
Total load (for W)			43.6 k		50 k

tions of the walls, which are indeed mostly solid because the openings are generally in the lower portion of the walls.

Using the value for W from the table, the total seismic load to the roof diaphragm in the north–south direction is thus

$$V = 0.1862(50) = 9.31 \text{ k}$$

which is only slightly higher than the value of 8 k as found for the north–south wind load in Chapter 6. As in that example, the roof plywood nailing and the edge chord members would be designed for this total load, assuming the diaphragm action to be as illustrated in Figure 6.3.

The end shear walls must be designed for the combined seismic and gravity loads as shown in Figure 7.2. Because the wall dead load was not included in the roof load tubulation for this direction, its lateral effect should be added as shown in the illustration, although the net effect is minor in this example. The gravity load of 6000 lb shown on the wall consists partly of roof load, assuming

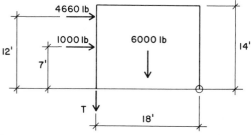

FIGURE 7.2   Overturn analysis of the end shear wall (Example 1).

some bearing wall function. We assume the wall itself to weigh 5000 lb and that its lateral load is 0.1862(5000) = 931 lb; say, 1000 lb. With the loads shown, the analysis of the wall is as follows:

Total lateral load: 5660 lb

Maximum shear stress: $\dfrac{5660}{18}$ = 314 lb/ft

Overturning M: (4.66)(12)(1.5)	= 83.9 k-ft
(1.0)(7)(1.5)	= 10.5
Total	= 94.4 k-ft
Dead load M: (6)(9)	= 54.0
Moment required for hold down:	= 40.4 k-ft

Required $T = \dfrac{40,400}{18}$ = 2244 lb

Section 2312(e)5 of the *UBC* requires that the system be designed for torsional effect assuming a minimum eccentricity of 5% of the gross dimension of the building. This is a questionable necessity for this symmetrical building and the relatively flexible wood diaphragm. In any event, the additional stress is minor, as we show. Using the value for the torsional moment of inertia as determined in Chapter 6 (see Figure 6.12 and Table 6.1.), the added stress in the end shear wall is

$$v = \frac{Tc}{J} = \frac{(9310)(0.05 \times 50)(25)}{32023} = 18 \text{ lb/ft}$$

If this added stress is also added to the overturn effect, the tie-down force would be increased as follows:

Added overturning M: (18 lb/ft)(18 ft)(12)(1.5) = 5832 lb-ft

Total new overturn M: 94.4 + 5.8 = 100.2 k-ft

New moment required for hold down: 100.2 − 54 = 46.2 k-ft

New required $T = \dfrac{46,200}{18}$ = 2567 lb

Referring to Table 7.1 for the value of W, the seismic load in the east–west direction is

$$V = 0.1862(43.6) = 8.12 \text{ k}$$

which is considerably higher than the wind load for this direction as found in Example 1 in Chapter 6. Ignoring torsion, the average stress in the shear walls is

$$v = \frac{8120}{57.33} = 142 \text{ lb/ft}$$

Assuming the dead loads to be approximately centered on the building, the added torsional stress on the front wall is

$$v = \frac{Tc}{J} = \frac{(8120)(7.67)(22.67)}{32023} = 44 \text{ lb/ft}$$

The front wall stress is thus a total of $142 + 44 = 186$ lb/ft. As usual, the negative torsional stress is not deducted from the stress on the rear wall. As discussed in Chapter 6, if the torsional effect is ignored and the load distribution is assumed to be on a peripheral basis, the analysis of the front wall is

$$\text{Total } V \text{ on the front wall} = \frac{1}{2}(8120) = 4060 \text{ lb}$$

$$v = \frac{4060}{14} = 290 \text{ lb/ft}$$

For a conservative design we would use the load of 142 lb/ft for the rear wall and 290 lb/ft for the front wall, covering all possibilities.

There are many additional design considerations to be kept in mind for the complete design of this building for seismic effects. Some of these are the horizontal sliding of the shear walls, the anchorage of the canopy, the combined stresses on the wall studs, and the cantilever action of the parapet. Many of these effects were illustrated in Example 1 in Chapter 6, and because the procedures are mostly similar once the lateral load magnitudes are determined, we refer the reader to that example for discussion of these problems.

Comparison of the wind and seismic analyses for the roof and wall diaphragms shows that the principal net difference between the two load considerations in this example building are the addition of tie downs for the front wall and the short rear walls.

7.2 Example 2

This example is the same as Example 3 in Chapter 6. It has the same building form as in the previous example, but it has a steel-framed roof and masonry walls. Assuming that the roof dead load is approximately the same and that the ceiling, canopy, and HVAC units are similar, the only difference between this example and the preceding one is the dead load of the exterior walls. The calculations for W are therefore the same as in Table 7.1, except for the exterior wall loads, which will be as follows:

East–west load:
  Wall DL to roof: (60 ft)(8 ft)(60 psf) = 28,800 lb
  Total W = 62,800 lb
  $V = 0.1862(62.8) = 11.7$ k

versus 4.8 k for wind and 8.12 k for the seismic load on the wood structure.

North–south direction:
  Wall DL to roof: (100 ft)(8 ft)(60 psf) = 48,000 lb
  Total W = 82,000 lb
  $V = 0.1862(82) = 15.27$ k

versus 8.0 k for wind and 9.31 k for the seismic load on the wood structure.

Although these loads are significantly higher than those for wind, the stresses in the roof deck and the shear walls are still well below critical limits for the steel deck and for the concrete block walls. Thus, the details of the construction are likely to be the same as for the wind design in Example 3 of Chapter 6.

Special note should be made of the requirements of Sections

2310, 2312(j)2D, and 2312(j)3A of the *UBC* with regard to the use of horizontal structures in conjunction with masonry or concrete walls. Section 2310 calls for a positive connection with a minimum capacity of 200 lb/ft. This requirement should be compared with the calculated load of the wall seismic force perpendicular to its plane.

Section 2312(g) requires that parts of structures be designed for a lateral force of

$$F_p = Z I C_p W_p$$

in which C_p is taken from Table 23-J of the *UBC* for various situations.

Using Z and I as previously determined and a C_p of 0.3 for the wall as required in the table, the wall load, as shown in Figure 7.3, thus becomes

$$F_p = (1.0)(1.0)(0.3) \ W_p = 0.3 \ W_p$$

For the wall alone, at 60 psf dead load, the total force is

$$F_p = 0.3(14 \times 60) = 252 \ \text{lb/ft of wall length}$$

and the T force required at the top of the wall is

$$T = \frac{252 \times 7}{12} = 147 \ \text{lb/ft of roof edge}$$

On the front wall, the weight of the canopy must be added to this load. Although Table 23-J of the *UBC* requires a C_p of 0.8 for the design of the canopy itself, in this case it is basically only

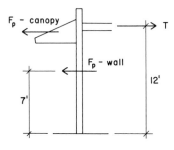

FIGURE 7.3 Outward lateral load on front wall (Example 1).

added weight to the wall. We thus consider only the addition of a load of

$$F_p = 0.3W_p = 0.3(100 \text{ lb/ft}) = 30 \text{ lb/ft}$$

With the canopy located approximately at the roof level, this force should be added entirely to the T at the top of the wall, making a total of

$$T = 147 + 30 = 177 \text{ lb/ft}$$

which is still less than the code minimum value of 200 lb/ft, as described previously. Thus, the calculated load is not critical in this example.

Design for the seismic force in the east–west direction is similar to that for wind in Example 3 in Chapter 6. The distribution of forces to individual piers is on the basis of their relative stiffnesses, and the eccentricity of the load from the center of stiffness produces a torsional rotation on the building.

7.3 Example 3

This example is the same as Example 10 in Chapter 6. See Figures 6.49 and 6.50 for the building plan and section. We will investigate the structure for the north–south seismic forces on the two-story-high shear walls.

Using *UBC* Formula 12–3A from Section 2312(d) we determine an approximate value for T as

$$T = \frac{0.05 \, h_n}{\sqrt{D}} = \frac{0.05(20)}{\sqrt{30}} = 0.183 \text{ sec}$$

This produces a calculated value for C of

$$C = \frac{1}{15 \sqrt{T}} = \frac{1}{15 \sqrt{0.183}} = 0.156$$

Because this is in excess of the maximum required value of 0.12 for C, we use the same factor for the total lateral force as for the one-story box system building. The total dead loads for the lateral forces applied through the roof and second floor diaphragms are

TABLE 7.2   Dead Load for the North–South Seismic Force: Example 3

Level	Source of load	Unit load	Load
Roof	Roof & ceiling	12 psf	$30 \times 40 \times 12 = 14.4$ k
	Exterior walls	15 psf	$5 \times 80 \times 15 \ = \ 6.0$
	Interior walls	10 psf	$5 \times 60 \times 10 \ = \ 3.0$
	HVAC unit	8 k	8.0
	Total		$= 31.4$ k
Second floor	Floor	10 psf	$30 \times 40 \times 10 = 12.0$
	Exterior walls	15 psf	$10 \times 80 \times 15 = 12.0$
	Interior walls	10 psf	$10 \times 60 \times 10 = \ 6.0$
	Total		$= 30.0$ k
	Total for base shear:		61.4 k

shown in Table 7.2. The total base shear for the two shear walls is thus

$$V = 0.1862 \ W = 0.1862(61.4) = 11.43 \text{ k}$$

This total force must be distributed to the roof and second floor in accordance with the requirements of Section 2312(e) of the *UBC*. The force at each level, F_x, is determined from Formula 12–7 as follows:

$$F_x = \frac{(V)(w_x h_x)}{\sum\limits_{i=1}^{n} w_i h_i}$$

in which

F_x is the force to be applied at each level (x)

w_x is the total dead load at level x

h_x is the height of level x above the base of the structure

(Notice that F_t has been omitted from the formula because T is less than 0.7 sec.)

The determination of the F_x values for this example is shown in Table 7.3. Using these F_x values, the shear wall loading and the

TABLE 7.3   Seismic Loads: Example 3

Level	w_x (kips)	h_x (ft)	$w_x h_x$	F_x (kips)
Roof	31.4	20	628	7.73
Second floor	30.0	10	300	3.70

$$\sum_{i=1}^{n} w_i h_i = \qquad\qquad 928$$

$$F_x = V \left[\frac{w_x h_x}{\displaystyle\sum_{i=1}^{n} w_i h_i} \right] = 11.43 \left[\frac{w_x h_x}{928} \right]$$

shear and moment diagrams are shown in Figure 7.4. The maximum shear stresses at the two levels are

Second story: $v = \dfrac{3865}{18} = 215$ lb/ft

First story: $v = \dfrac{5715}{18} = 318$ lb/ft

both of which are within the load range for a plywood wall.

Figure 7.5 shows the loadings for the overturn analyses of the two stories. The assumed dead loads shown reflect the use of the walls as bearings walls and thus include some of the roof and

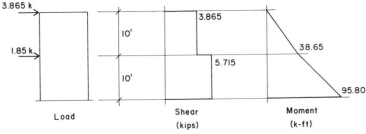

FIGURE 7.4   Analysis of the two-story shear wall for seismic load (Example 3).

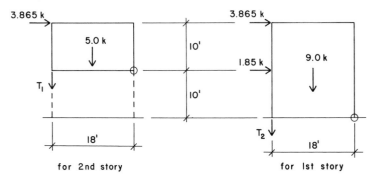

for 2nd story   for 1st story

FIGURE 7.5 Overturn analysis of the shear walls (example 3).

second-floor loads. For the second story the overturn analysis is as follows:

Overturn M: $(3.865)(10)(1.5\ SF) = 58.0$ k-ft
Dead load M: $(5.0)(9)$ $= \underline{45.0}$

Moment required for hold down: 13.0 k-ft

Required $T_1 = \dfrac{13,000}{18} = 722$ lb

which is a very small tie-down force for the wood wall.
    For the first story wall, the overturn analysis is as follows:

Overturn M: $(3.865)(20)(1.5\ SF) = 116.0$ k-ft
              $(1.85)(10)(1.5)$ $= \underline{27.8}$
        Total $= 143.8$ k-ft
Dead load M: $(9.0)(9)$ $= \underline{81.0}$

Moment required for hold down: 62.8 k-ft

Required $T_2 = \dfrac{62,800}{18} = 3489$ lb

This is a significant hold-down force, but it is within the range of ordinary hold-down devices (see the Appendix), assuming that the foundations can develop the resistance required.

A comparison of these results with those found in the wind analysis in Example 10 in Chapter 6 shows that there is very little difference between the wind and earthquake effects on this building. The details of the shear wall construction are essentially the same for the two loading conditions.

7.4   Example 4

This example is the same as Example 11 in Chapter 6. The building is the same as in the preceding example, except that the roof and floor structure is steel and the shear walls are of reinforced concrete block. Although this change does not affect the wind loads, the added weight of the masonry walls and the concrete fill on the floor deck result in higher seismic forces. The calculation of dead loads for the north–south seismic force is shown in Table 7.4. We assume that the remainder of the exterior walls are also of concrete block, with an average weight of 40 psf to account for the wall openings. For the shear wall we assume a nominal 8-in. wall with an average weight of 60 psf.

The total load found in Table 7.4 is that of the roof and second-floor diaphragm tributary loads plus the full weight of the shear walls. It thus represents the full base shear load for the

TABLE 7.4   Dead Load for the North–South Seismic Force: Example 4

Level	Source of Load	Unit load	Load	
Roof	Roof and ceiling	12 psf	$30 \times 40 \times 12 =$	14.4 k
	Exterior walls	40 psf	$5 \times 80 \times 40 =$	16.0
	Partitions	10 psf	$5 \times 60 \times 10 =$	3.0
	Shear walls	60 psf	$10 \times 36 \times 60 =$	21.6
	HVAC unit	8 k		8.0
			Total $=$	63.0 k
Second floor	Floor	40 psf	$30 \times 40 \times 40 =$	48.0
	Exterior walls	40 psf	$10 \times 80 \times 40 =$	32.0
	Partitions	10 psf	$10 \times 60 \times 10 =$	6.0
	Shear walls	60 psf	$10 \times 36 \times 60 =$	21.6
			Total $=$	107.6 k
Total load for base shear:				170.6 k

TABLE 7.5 Seismic Loads: Example 4

Level	w_x (in kips)	h_x (in feet)	$w_x h_x$	F_x (in kips)
Roof	63.0	20	1260	17.14
Second floor	107.6	10	1076	14.63

$\sum_{i=1}^{n} w_i h_i =$ 2336

$$F_x = V \left[\dfrac{w_x h_x}{\displaystyle\sum_{i=1}^{n} w_i h_i} \right] = 31.77 \left[\dfrac{w_x h_x}{2336} \right]$$

shear walls. We thus calculate the base shear as

$$V = 0.1862W = 0.1862(170.6) = 31.77 \text{ k}$$

As in the preceding example, this shear must be distributed to the roof and second-floor levels according to Section 2312(e) of the *UBC*. Table 7.5 gives the calculation of the F_x values for the two levels. The shear and moment diagrams for this loading on the two-story wall are shown in Figure 7.6. The maximum shear stresses at the two levels, assuming a 45% solid block, are

Second-story wall:

Load per ft of wall: $v = \dfrac{8570}{18} = 476 \text{ lb/ft}$

Stress on net wall: $v = \dfrac{476}{(12)(7.625)(0.45)} = 11.6 \text{ psi}$

Table 24–H of the *UBC* (see the Appendix) gives the lowest value of shear as 35 psi, so that this stress condition is not critical. Notice that a footnote to Table 24–H requires a 50% increase in the load for calculation of the shear stress. This still will not create a critical condition in this example.

At the first story, the net stress can be proportioned from the shear forces as

$$v = \dfrac{15,890}{8570} (11.6) = 21.5 \text{ psi}$$

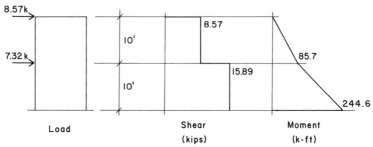

FIGURE 7.6   Analysis of the two-story shear wall for seismic load (Example 4).

With the required increase of 50%, this result jumps to 32.2 psi, but it is still not a critical stress condition. Thus, the 8-in. block walls are adequate for the shear force, although the reinforcing must be designed to take the shear stresses.

Figure 7.7 shows the loadings for the overturn analyses of the two stories. As in the previous example, the wall dead load includes parts of the roof and second-floor dead loads. For the second story the overturn analysis is as follows:

Overturn M: (8.57)(10)(1.5 SF) = 128.6 k-ft
Dead load M: (12.0)(9)   = 108.0

Moment required for hold down: 20.6 k-ft

$$\text{Required } T_1 = \frac{20,600}{18} = 1144 \text{ lb}$$

which can easily be developed by the doweling of the wall reinforcing.

For the first-story wall the overturn analysis is as follows:

Overturn M: (8.57)(20)(1.5 SF) = 257.1 k-ft
                      (7.32)(10)(1.5)   = 109.8
                      Total   = 366.9 k-ft
Dead load M: (28.0)(9)   = 252.0
Moment required for hold down: = 114.9 k-ft

$$\text{Required } T_2 = \frac{114,900}{18} = 6383 \text{ lb}$$

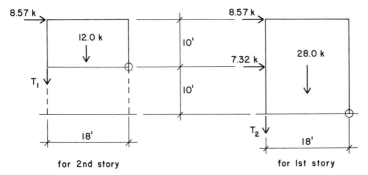

FIGURE 7.7   Overturn analysis of the shear walls (Example 4).

This is a significant force, but it can probably be developed by the wall reinforcing because the end of the wall would be designed as a column for both the gravity and shear wall designs. The design for compression due to the seismic overturn moment and the gravity load would most likely govern the design of the wall ends.

Comparison of the results found in this example with those found in Example 11 in Chapter 6 show the seismic effect to be much higher. However, the stress analysis and overturn investigation indicate that the construction would most likely be the same for both loading conditions. An additional design problem is that of the shear wall foundation, which must function for both gravity and lateral loads. Because of the higher load magnitudes in the next example, we discuss this problem in Example 5. An illustration of the use of *UBC* Formula 12–9 from Section 2312(j) to find the design loads for the diaphragms and collectors at each level is also given in Example 5.

7.5   Example 5

This example is the same as Example 12 in Chapter 6. See Figure 6.57 for the building plan and section. For the seismic analysis we assume the following:

$Z = 1.0, I = 1.0, S$ not determined from T_s.

$K = 1.33$ for the shear wall braced structure.

Dead loads:

Roof and ceiling: 12 psf

Upper floors: 40 psf

Exterior walls: average of 15 psf

Interior partitions: 10 psf

Shear wall: assume 80 psf

Roof top HVAC equipment: assume total of 15 k

Section 2312(d) of the *UBC* permits a calculation of the value of *T* using the simplified formula

$$T = \frac{0.05\, h_n}{\sqrt{D}} = \frac{0.05(50)}{\sqrt{90}} = 0.263 \text{ sec}$$

which results in a calculated value for *C* of

$$C = \frac{1}{15\sqrt{T}} = \frac{1}{15\sqrt{0.263}} = 0.13$$

Because this result is in excess of the code maximum value of 0.12, we will use the maximum CS value of 0.14 as before.

Section 2312(e) of the *UBC* requires a concentrated force at the top of a structure, but only when *T* exceeds 0.7 seconds. With this F_t force eliminated, the lateral forces at each level are determined from Formula 12–7 of the *UBC* as follows:

$$F_x = \frac{(V)(w_x h_x)}{\sum\limits_{i=1}^{n} w_i h_i}$$

in which

F_x is the lateral force to be applied at each level (*x*)

w_x is the total dead load at level *x*

h_x is the height of level *x* above the base of the structure

The tabulation of the building dead loads for the determination of the values of w_x and the total lateral force, *V*, are shown in Table 7.6. Using these values, the determination of the values for

TABLE 7.6 Dead Load for the North–South Seismic Force: Example 5

Level	Source of load	Unit load (psf)	Load		
Roof	Roof and ceiling	12	$120 \times 90 \times 12$	=	129.6 k
	Exterior walls	15	$420 \times 7.5 \times 15$	=	47.2
	Interior walls	10	$300 \times 7.5 \times 10$	=	22.5
	Shear walls	80	$60 \times 7.5 \times 80$	=	36.0
	HVAC units			=	15.0
			Subtotal	=	250.3 k
Third floor	Floor	40	$120 \times 90 \times 40$	=	432.0
	Exterior walls	15	$420 \times 15 \times 15$	=	94.5
	Interior walls	10	$300 \times 15 \times 10$	=	45.0
	Shear walls	80	$60 \times 15 \times 80$	=	72.0
			Subtotal	=	643.5 k
Second floor	Floor	40	$120 \times 90 \times 40$	=	432.0
	Exterior walls	15	$420 \times 17.5 \times 15$	=	110.2
	Interior walls	10	$300 \times 17.5 \times 10$	=	52.5
	Shear walls	80	$60 \times 17.5 \times 80$	=	84.0
			Subtotal	=	678.7 k
First floor	Shear walls	80	$60 \times 10 \times 80$	=	48.0
	(Assume remainder of first story load to first floor)				
Total for V at first floor level:					1620.5 k
Foundation					100.0
Total for V at bottom of footing:					1720.5 k

For F_x: $V = 0.1862(1720.5) = 320$ k

F_x at each level is as shown in Table 7.7. Notice that we are assuming the foundation to be without a basement, as illustrated in the second case for Example 12 in Chapter 6. Thus, the distances used for h_x are measured from the bottom of the footing. The weight of the foundation is included in the determination of the total value for V, but it has not otherwise been included in the determination of the F_x values because its moment effect is quite small.

Using the F_x values determined in Table 7.7, the loading and the shear and moment diagrams for the shear wall are as shown in

TABLE 7.7 Seismic Loads: Example 5

Level	w_x (kips)	h_x (ft)	$w_x h_x$	F_x (kips)
Roof	250.3	58	14,517	75.4
Third floor	643.5	43	27,671	143.8
Second floor	678.7	28	19,004	98.8
First floor	48	8	384	2.0
$\sum_{i=1}^{n} w_i h_i =$			61,576	

$$F_x = 320 \left(\frac{w_x h_x}{61,576} \right)$$

Figure 7.8. Notice that the F_x values have been divided by two for the load on a single shear wall.

At the third story the shear force of 37.7 k produces a unit shear on the wall of

$$v = \frac{37,700}{30} = 1257 \text{ lb/ft}$$

Assuming an 8-in. block wall with 45% solid block, the stress on the net wall section is

$$v = \frac{1257}{(0.45)(12)(7.625)} = 30.5 \text{ psi}$$

This is not a critical stress for the reinforced concrete block wall. (See *UBC* Table 24–H in the Appendix.) A footnote to *UBC* Table 24–H requires that the shear load be increased by 50% for shear stress calculation. Applying this increase, together with the allowable one-third increase due to seismic load, the actual design shear stress for the wall is:

$$v = (1.5) \frac{30.5}{1.33} = 34.4 \text{ psi}$$

which is still below the lowest allowable stress in the table.

The ends of the shear wall would most likely be designed as columns to carry the roof framing. If this is done, the end column

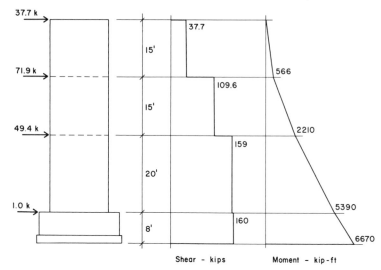

FIGURE 7.8   Analysis of the three-story shear wall for seismic load (Example 5).

forces due to the seismic overturn moment can probably be easily accommodated. If a tie-down force is required, the doweling of the reinforcing will provide it.

At the second story the shear force of 109.6 k produces a unit shear on the wall of

$$v = \frac{109,600}{30} = 3653 \text{ !b/ft}$$

which is high for an 8-in. block wall.

Assuming a 12-in. block wall with 45% solid blocks, the stress on the net section is

$$v = \frac{3653}{(0.45)(12)(11.625)} = 58 \text{ psi}$$

With the adjustments discussed previously, the actual design stress will be slightly higher, but it is still within the conceivable range of the masonry wall. Overturn moments can also probably be handled with the use of reinforced masonry columns at the ends of the walls.

At the first story the seismic shear and moment are probably beyond the range of a feasible masonry wall. An alternative would be to use a reinforced concrete wall with end columns. The unit shear on the wall is

$$v = \frac{159{,}000}{30} = 5300 \text{ lb/ft}$$

With a 10-in.-thick concrete wall the unit stress in the wall is

$$v = \frac{5300}{(10)(12)} = 44 \text{ psi}$$

which is quite low for reinforced concrete.

If the wall foundation is assumed to be the same as for the second case in Example 12 in Chapter 6, the loading condition for the overturn analysis on the footing is as shown in Figure 7.9.

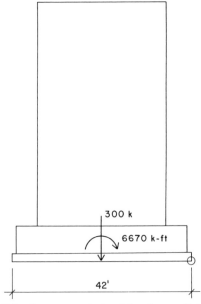

300 k

6670 k-ft

42'

FIGURE 7.9   Overturn analysis of the shear wall (Example 5).

Because the shear walls in this example are slightly heavier than those in the Chapter 6 example, we assume a total dead load of 300 k. The overturn analysis is thus as follows:

Overturning M: (6670)(1.5 SF) = 10,005 k-ft
Dead load M: (300)(21)   = 6,300 k-ft

This analysis shows that the safety factor is not provided by the dead load alone. The options are, therefore, to increase the size of the wall foundation or to develop ties to other parts of the building foundation to provide additional resistance to the overturn.

Comparison of the seismic load on this example with the wind loads on the same building in Example 12 of Chapter 6 show that the seismic loading is considerably higher. The increases required in the foundation are considerable, making the lateral system solution questionable for the building if the seismic load must be considered. Although the structure may be adequate for wind, if the seismic load must be accommodated, it is advisable to reconsider the general design of the structure. Some possible options are the following:

1. Reduce the building dead load as much as possible to lower the lateral shear force.
2. Increase the length of the shear walls or add more shear walls.
3. Use an interactive shear wall and ductile rigid frame. The major gain that this will achieve is in the reduction of the K factor: from 1.33 to 0.80, a reduction of 40%.

The use of the F_x forces obtained from Formula 12–7 of the *UBC* as the lateral loads at the various levels is intended to bring the static analysis into closer agreement with true dynamic behavior in the multilevel structure. These forces are appropriate for use in the design of the vertical elements of the lateral resistive system but not for the design of the horizontal structures at the various levels. Formula 12–9 of the *UBC* is used to obtain the forces to be used for the design of the diaphragms and collectors at each level. The implementation of this formula is illustrated in

TABLE 7.8 Horizontal Diaphragm and Collector Loads: Example 5

Level	w_{px} (kips) A	$\sum_{1=x}^{n} F_1$ (kips) B	$\sum_{1=x}^{n} w_1$ (kips) C	$\dfrac{B}{C}$ D	F_{px} (kips) E
Roof	250.3	75.4	250.3	0.301	75.4
Third floor	643.5	219.2	893.8	0.245	157.7
Second floor	678.7	318.0	1572.5	0.202	137.1

$$F_{px} = \dfrac{\sum_{1=x}^{n} F_1}{\sum_{1=x}^{n} w_1} (w_{px}) \quad \text{or} \quad \dfrac{B}{C} (A)$$

Table 7.8. The entries in the table, labeled "A" through "E" at the top of the table, are determined as follows:

A is the same as w_x in Table 7.6.

B is a summation of the F_x values from Table 7.6.

C is a summation of the w_x values from Table 7.6.

D, as indicated, is the ratio of entry B to entry C.

E is obtained from *UBC* Formula 12–9, or with reference to the entries in Table 7.8.

$$E = \dfrac{B}{C} (A)$$

7.6   Example 6

This example is a modification of the building in the previous example. As shown in the plan for the first floor in Figure 7.10 and the sections in Figure 7.11, it is desired to eliminate the interior shear walls at the first story. Thus, the vertical resistive system at the first story becomes the four end shear walls.

In order to use this scheme, it would be necessary for the second-floor structure to achieve the transfer of the horizontal

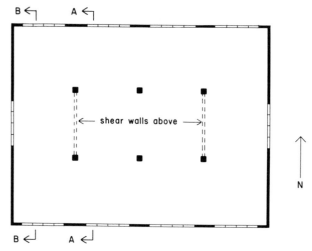

FIGURE 7.10   Building plan (Example 6).

force from the interior walls to the exterior walls. The loading of the horizontal structure for the lateral loads is thus as shown in Figure 7.12. The direct load to the second-floor diaphragm is applied as a uniform load, and the loads from the upper stories become concentrated loads at the locations of the interior shear walls. This structure would most likely be either a reinforced concrete and beam system or a heavily braced steel frame.

Although the horizontal shear force is removed from the interior walls at the second floor, the overturn effect is not. It must be

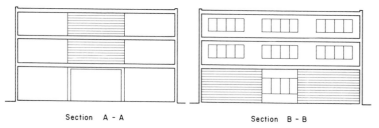

Section   A - A   Section   B - B

FIGURE 7.11   Building sections (Example 6).

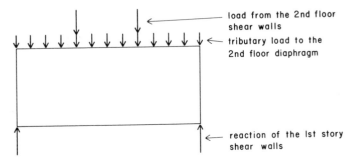

FIGURE 7.12 Function of the second-floor diaphragm (Example 6).

carried down to the foundations by extending the wall ends as columns at the first story, as shown in the section in Figure 7.11.

With the lateral loads approximately the same as in the previous example, it becomes feasible to consider the use of reinforced masonry for the first-story shear walls in this example. Their aggregate length is considerably more, and the unit shear stress would probably drop into the range of the masonry construction.

7.7 Example 7

This example consists of the same building used in Example 13 in Chapter 6. We will determine the seismic load to the roof diaphragm and the five rigid bents for comparison to the wind load. Assuming the exterior walls to span vertically and an absence of interior walls, the total weight for the roof diaphragm load is

Roof: (assumed 25 psf *DL*)(100 × 50) = 125,000 lb
Walls: (assumed 20 psf)(200)(10)   = 40,000

    Total   = 165,000 lb

Assuming the frame to be a ductile frame as qualified by the *UBC*, the *K* factor from *UBC* Table 23–I (see the Appendix) is 0.67. With *S* undetermined and using the maximum *CS* value of 0.14, the total lateral load to the roof is thus

$$V = ZIKCSW = (1.0)(1.0)(0.67)(0.14)W = 0.0938W$$

$$= 0.0938(165) = 15.48 \text{ k}$$

With the roof diaphragm assumed to be relatively rigid, as in the wind analysis in Chapter 6, this load will be divided equally between the five bents. Thus, the load per bent is

$$V = \frac{15.48}{5} = 3.1 \text{ k/bent}$$

Because this is less than the wind load as found in Chapter 6, it is not critical for the building.

7.8 Example 8

This example consists of the same building as used in Example 15 in Chapter 6. The lateral resistive structure consists of seven column/beam bents in the north–south direction. The dead loads for determination of the horizontal diaphragm loads are essentially the same as in Example 5 in this chapter, except for the absence of the interior shear walls. Therefore, we use 90% of the loads as tabulated for the various levels in Table 7.6.

The total lateral load on the structure is further reduced because of the lower K value for the rigid frame. The amount of this reduction is

$$\frac{K \text{ for the frame}}{K \text{ for the shear wall}} = \frac{0.67}{1.33} = 0.50 \text{ or } 50\%$$

The F_x values to be applied to the various levels, as required by Section 2312(e)1 of the *UBC* and determined from *UBC* Equation 12–7 may thus be approximately found by applying these reductions to the F_x values as found in Table 7.7 for Example 5. Because the F_x values in the table are for the entire building, we also divide by 7 to find the load per bent. Thus, the bent loadings are

$$\text{Roof: } (0.90)(0.50)(75.4)/7 = 4.85 \text{ k}$$

$$\text{Third floor: } (0.90)(0.50)(143.8)/7 = 9.24 \text{ k}$$

$$\text{Second floor: } (0.90)(0.50)(98.8)/7 = 6.35 \text{ k}$$

A comparison with the wind loading as shown in Figure 6.81 indicates that these loadings are slightly higher, and therefore the seismic loading is critical for the building.

7.9   Example 9

This example uses the same building as that used in Example 18 in Chapter 6. The vertical system for lateral resistance in the north–south direction consists of two trussed bents, as shown in Figure 6.88. The building is otherwise similar to that in Example 5 in this chapter, with the X-braced bents replacing the shear walls.

Because the required K factor is the same for shear walls and braced frames, the only difference in the lateral loads between this example and Example 5 is in the weight reduction caused by the elimination of the heavy shear walls. Thus, the total lateral forces for the various levels are approximately 90% of those determined in Table 7.5. A comparison of these values with those found for the wind loading in Chapter 6 shows that the seismic load is critical for the building, as it was also for the shear wall braced building in Example 5.

It should be noted that Section 2312(j)1G of the *UBC* requires that the members of a braced frame be designed for 1.25 times the load as determined from *UBC* Formula 12–1 ($V = ZIKCSW$) and that the connections for the members be designed for this increased load or the full strength of the members, whichever is higher. In addition, the usual one-third increase in allowable stress is not allowed for the connections.

7.10   Example 10

This example consists of the free-standing masonry wall that was designed for wind in Example 19 in Chapter 6. For seismic analysis the wall height must be measured from the top of the footing. With the shallow footing as designed in Chapter 6, this gives a wall of 6 ft 8 in. in height.

Table 23-J of the *UBC* requires a C_p value of 0.3 for masonry fences over 6 ft high. However, a footnote to the table allows the table values to be reduced by one-third when the structures are laterally supported only at the ground level. Using the wall weight of 400 lb, as found in Chapter 6, the lateral load on the wall is thus determined from *UBC* Formula 12–8 as follows:

$$F_p = ZIC_pW_p$$
$$= (1.0)(1.0)(0.2)(400) = 80 \text{ lb}$$

A comparison with the wind analysis in Chapter 6 shows this to be a less critical condition. The wall therefore can be designed for the wind load. However, the wind load of 20 psf used in Chapter 6 is quite high. If the reduction for fences, as provided by Section 2311(h) of the *UBC* is applied to the lowest wind zone load from *UBC* Table 23–F, the design wind pressure for the wall is only 10 psf, and the seismic load governs the design.

7.11   Example 11

This example consists of the tall, round stack that was analyzed for wind in Example 20 in Chapter 6. Table 23–J of the *UBC* gives a value of 0.3 for C_p for this structure, which can be reduced to 0.2 with the provision of the footnote to the table. Thus, assuming the dead load of the stack to be 60 k as was used in Chapter 6, the total lateral load on the stack from *UBC* Formula 12–8 is

$$F_p = ZIC_pW_p = (1.0)(1.0)(0.2)(60) = 12 \text{ k}$$

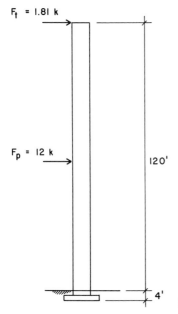

F_t = 1.81 k

F_p = 12 k

120'

4'

FIGURE 7.13   Seismic load (Example 11).

This load is slightly less than the wind load as found in Chapter 6. However, the fundamental period of this slender structure will likely be quite long, and an additional force must be applied at the top of the stack as required by Section 2312(e) and Formula 12–6 of the *UBC*. Although the true period should be determined from a dynamic analysis of the stack structure, an approximation can be made using *UBC* Formula 12–3A. This formula is intended only for rectangular building forms, so the value found is only approximate for this tall, round structure. Using the formula gives

$$T = \frac{0.05\, h_n}{\sqrt{D}} = \frac{(0.05)(122)}{\sqrt{8}} = 2.16 \text{ sec}$$

This is considerably higher than the minimum of 0.7 sec, which is the limit for the use of F_t in Section 2312(e). With the approximate value for T, the required F_t is

$$F_t = 0.07\, TV = (0.07)(2.16)(12) = 1.81 \text{ k}$$

With this load added to the lateral force previously found, the stack load is as shown in Figure 7.13. The overturning moment at the toe of the footing is thus a total of

$$
\begin{aligned}
M = (1.81)(124) &= 224 \text{ k-ft}\\
(12)(63) &= \underline{756}\\
\text{Total} &= 980 \text{ k-ft}
\end{aligned}
$$

which is only slightly higher than the total wind overturning moment as found in Table 6.3. However, a more accurate determination of the period for the stack may result in a larger value of F_t and a higher overturning moment.

8

Special Problems

||||||||||||||||||||||||||||||||||||||||||||||||||||||

The material in this chapter deals with some of the special problems that must be considered in the design for lateral loads. These
problems have been discussed in other chapters and in most cases
have been encountered in the example problems in Chapters 6
and 7. The purpose here is to present a more generalized discussion of the specific problems.

8.1 Deflections Caused by Lateral Loads

The deflection of the lateral resistive structural system is of concern for a number of possible reasons. Some of these are as
follows:

The potential effect of the movements and deformations on
other parts of the building construction, such as windows and
partitions.

The need for determination of relative stiffnesses of elements
as a basis for finding the distribution of loads to parts of the
system.

The need to determine actual dimensions of movement to be
provided for in separation joints where they are required.

For these reasons, and for any others that may be critical to the building design, we will discuss the deflection of four types of elements: horizontal diaphragms, vertical shear walls, X-braced frames, and rigid frames.

Horizontal Diaphragms. As spanning elements, the relative stiffness and actual dimensions of deformation of horizontal diaphragms depends on a number of factors, such as:

The materials of the construction.

The continuity of the spanning diaphragm over a number of supports.

The span-to-depth ratio of the diaphragm.

The effect of various special conditions, such as chord length changes, yielding of connections, influence of large openings, and so on.

In general, wood and light gauge metal decks tend to produce quite flexible diaphragms whereas poured concrete decks tend to produce the most rigid diaphragms. Ranging between these extremes are decks of lightweight concrete, gypsum concrete, and composite constructions of lightweight concrete fill on metal deck. For true dynamic analysis, the variations are more complex because the weight and degree of elasticity of the materials must also be considered.

With respect to their span-to-depth ratios, most horizontal diaphragms approach the classification of deep beams. As shown in Figure 8.1, even the shallowest of diaphragms, such as the maximum 4 to 1 case allowed for a plywood deck by the *UBC*, tends to present a fairly stiff flexural member. As the span-to-depth ratio falls below about 2, the deformation characteristic of the diaphragm approaches that of a deep beam, in which the deflection is primarily caused by shear strain rather than by flexural strain. Thus, the usual formulas for deflection caused by flexural strain become of limited use.

The following formula (see *Western Woods Use Book,* Ref. 8) is

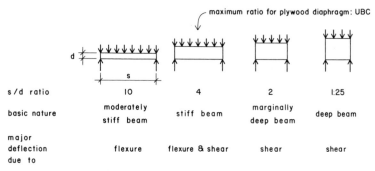

FIGURE 8.1 Behavior of horizontal diaphragms related to depth-to-span ratios.

used for the calculation of deflection of simple span plywood diaphragms:

$$\Delta = \frac{5\,vL^3}{8\,EAb} + \frac{vL}{4\,Gt} + 0.094\,Le_n + \Sigma\,\frac{(\Delta_c X)}{2b}$$

in which the four terms account for four different contributions to the deflection, as follows:

Term 1 accounts for the length change of the chords.

Term 2 accounts for the shear strain in the plywood panels.

Term 3 accounts for the lateral bending of the nails.

Term 4 accounts for additional change in the chord lengths caused by slip in the chord splices.

The reader is referred to the Ref. 8 or to other sources on wood design for data and explanations of the factors in this equation and for examples of its use.

The deflection of steel deck diaphragms is discussed and illustrated in the *Inryco Lateral Diaphragm Data Manual 20-2* (Ref. 9). The formula used for calculating the deflection of a simple span deck is

$$\Delta_t = \frac{5}{384}\,\frac{W \times L_s^4 \times 1728}{E \times I} + q \times F \times \frac{L_s}{2} \times 10^{-6}$$

in which the first term accounts for flexural deflection caused by the length change of the chords and the second term for shear strain and panel distortion in the diaphragm web.

The quantities q and F vary as a function of the type and gauge of the deck, the fastening patterns and methods used, and the possible inclusion of concrete fill. Data for a wide range of the possible combinations of these variables, using the products of the Inland Steel Company, are given in Ref. 9. Also given is a method and the necessary data for determination of the relative stiffness of the diaphragm, which is an important issue in determining the proper method for distribution of forces to the vertical resistive elements of the lateral bracing system.

Vertical Shear Walls. As with the horizontal diaphragm, there are several potential factors to consider in the deflection of a shear wall. As shown in Figure 8.2, shear walls also tend to be relatively stiff in most cases, falling into deep beam instead of ordinary flexural members.

As discussed in Chapter 5 and illustrated in the examples in Chapter 6, there are two general cases for the vertical shear wall: the cantilever and the doubly fixed pier. The cantilever, fixed at its base, is the most commonly used. Fixity at both the top and bottom of the wall usually affects deflection only when the wall is

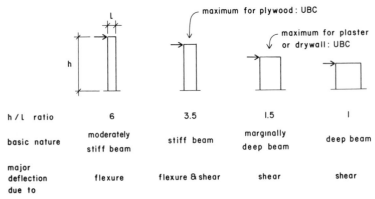

FIGURE 8.2 Behavior of vertically cantilevered elements related to height-to-length ratios.

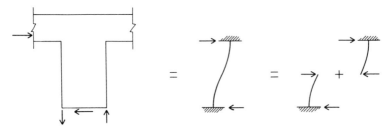

FIGURE 8.3 Deflection assumptions for a fixed pier.

relatively short in length with respect to its height. Walls with long lengths in proportion to their height fall into the deep beam category, in which the predominant shear strain is not affected by the fixity of the supports.

As shown in Figure 8.3, if the doubly fixed pier is assumed to have an inflection point at its mid-height, its deflection can be approximated by considering it to be the sum of the deflections of two half-height cantilevered piers. Yielding of the supports and flexure in the horizontal structure will produce some rotation of the assumedly fixed ends, which will result in some additional deflection.

The following formula (Ref. 8) is used for the calculation of deflection of cantilevered plywood shear walls, similar to that for the plywood horizontal diaphragm:

$$\Delta = \frac{8vh^3}{EAb} + \frac{vh}{Gt} + 0.376\,he_n + d_a$$

in which the four terms account for the following:

Term 1 accounts for the change in length of the chords (wall end framing).

Term 2 accounts for the shear strain in the plywood panels.

Term 3 accounts for the nail deformation.

Term 4 is a general term for including the effects of yield of the anchorage.

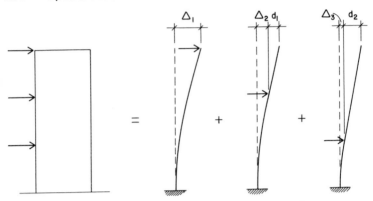

FIGURE 8.4   Deflection of a multistory shear wall.

As before, the reader is referred to Ref. 8 for the explanation and examples of use of the formula. The formula can also be used for the calculation of the deflection of a multistory wall, as shown in Figure 8.4. For the loading as shown in the illustration, a separate calculation would be made for each of the three loads (Δ_1, Δ_2, and Δ_3). To these would be added the deflection at the top of the wall caused by the rotation effects of the lower loads (d_1 and d_2). Thus, the total deflection at the top of the wall would be the sum of the five increments of deflection.

Rotation caused by soil deformation at the base of the wall can also contribute to the deflection of shear walls (see Figure 8.5). This is especially critical for tall walls on isolated foundations placed on relatively compressible soils, such as loose sand and soft clay, a situation to be avoided if at all possible. The design of such a foundation is discussed and illustrated in Example 12 in Chapter 6.

X-braced Frames. The deflection of single-story X-braced frames is usually caused primarily by the tension elongation of the diagonal X members. As shown in Figure 8.6, this deformation tends to move the rectangular bays of the frame into a parallelogram form. The approximate value of this deflection, labeled "d" in Figure 8.6, can be derived as follows.

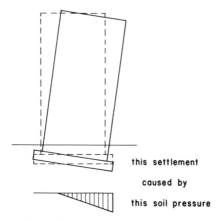

FIGURE 8.5   Shear wall drift caused by settlement.

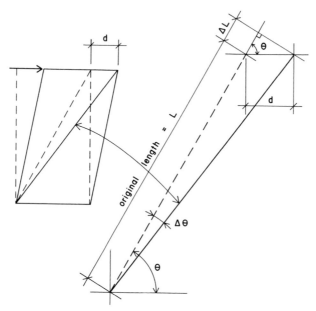

FIGURE 8.6   Drift of an X-braced frame.

Assuming the change in the angle of the diagonal, $\Delta\Theta$ in the figure, to be quite small, the change in length of the diagonal may be used to approximate one side of the triangle of which "d" is the hypotenuse. Thus:

$$d = \frac{\Delta L}{\cos \Theta} = \frac{T/AE}{\cos \Theta} = \frac{T}{AE \cos \Theta}$$

in which: T = the tension in the X caused by the lateral load
A = the cross-sectional area of the X
E = the elastic modulus of the X
Θ = the angle of the X from the horizontal

The deflection of multistory X-braced frames has two components, both of which may be significant. As shown in Figure 8.7, the first effect is caused by the elongation of the diagonal X, as discussed for the single-story frame. These deflections occur in each level of the frame and can be calculated individually and summed up for the whole frame. The second effect is caused by the change in length of the vertical members of the frame as a result of the overturning moment. Although this effect is also present in the single-story frame, it becomes more pronounced as the frame gets taller with respect to its width. These deflections of the cantilever beam can be calculated using standard formulas,

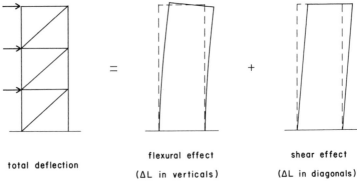

total deflection

flexural effect
(ΔL in verticals)

shear effect
(ΔL in diagonals)

FIGURE 8.7   Deflection of a multistory X-braced frame.

such as those given for Cases 21 and 22 in the Beam Diagrams and Formulas in Section 2 of the *Manual of Steel Construction* (Ref. 5).

As for the tall shear wall, the possibility of rotation of the foundation should be investigated for the multistory X-braced frame.

For horizontal structures consisting of X-braced systems, such as those shown in Example 17 in Chapter 6, deflections can be found by the usual methods of determination for spanning trusses.

Rigid Frames. The general calculation of lateral deflections for rigid frames is well beyond the scope of this book. The relatively simple case of the single-span bent is illustrated and discussed in Example 13 in Chapter 6. For a general treatment of the deflection of multistory rigid frames, the reader may refer to *High-Rise Building Structures* (Ref 15).

It should be borne in mind, as has been discussed previously, that the so-called "rigid" frame is actually quite flexible with regard to lateral deflections. This is a primary reason for the assignment of the low value for K in the equation for seismic load in the *UBC*.

8.2 Separation Joints

During the swaying motions induced by earthquakes, different parts of a building tend to move independently because of the differences in their masses, their fundamental periods, and variations in damping, support constraint, and so on. With regard to the building structure, it is usually desirable to tie it together so that it moves as a whole as much as possible. Sometimes, however, it is better to separate parts from one another in a manner that permits them a reasonable freedom of motion with respect to one another.

Figure 8.8 shows some building forms in which the extreme difference of period of adjacent masses of the building make it preferable to cause a separation. Designing the building connection at these intersections must be done with regard to the specific

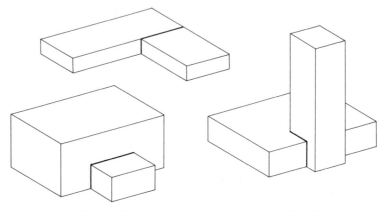

FIGURE 8.8   Potential situations requiring separation joints.

situation in each case. Some of the considerations to be made in this are discussed in the following paragraphs.

The Specific Direction of Movements.   In generally rectangular building forms, such as those shown in the examples in Figure 8.8, the primary movements are in the direction of the major axes of the adjacent masses. Thus, the joint between the masses has two principle forms of motion: a shear effect parallel to the joint, and a together–apart motion perpendicular to the joint. In building forms of greater geometric complexity, the motions of the respective masses is more random, and the joint action is much more complex.

The Actual Dimensions of Movement at the Joint.   If the joint is to be truly effective and if the adjacent parts are not allowed to pound each other, the actual dimension of the movements must be safely tolerated by the separation. The more complex the motions of the separate masses, the more difficult it is to predict these dimensions accurately, calling for some conservative margin in the dimension of the separation provided.

Detailing the Joint For Effective Separation.   Because the idea of the joint is that structural separation is to be provided while still achieving the general connection of the adjacent parts, it is necessary to make a joint that performs both of these seemingly contradictory functions. Various techniques are possible using connections that employ sliding, rolling, rotating, swinging, or flexible elements that permit one type of connection while having a freedom of movement for certain directions or types of motion. The possibilities are endless, and the specific situation must be carefully analyzed in order to develop an effective and logical joint detail. In some cases the complexity of the motions, the extreme dimension of movement to be facilitated, or other considerations may make it necessary to have complete separation —that is, literally to build two separate buildings very close together.

Facilitating Other Functions of the Joint.   It is often necessary for the separation joint to provide for functions other than those of the seismic motions. Gravity load transfer may be required through the joint. Nonstructural functions, such as weather sealing, waterproofing, and the passage of wiring, piping, or ductwork through the joint, may be required. Figure 8.9 shows two typical cases in which the joint achieves structural separation while providing for a closing of the joint. The drawing on the left shows a flexible flashing or sealing strip used to achieve weather or water tightness of the joint. The drawing on the right symbolizes the usual solution for a floor, in which a flat element is attached to one side of the joint and is allowed to slip on the other side.

Figure 8.10 shows a number of situations in which partial structural separation is achieved. The details of such joints are often quite similar to those used for joints designed to provide

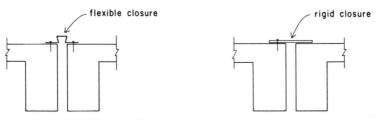

FIGURE 8.9   Closure of horizontal separation joints.

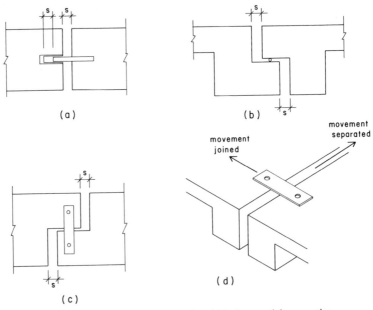

FIGURE 8.10 Optional means for achieving partial separation.

separation for thermal expansion. Detail (a) shows a key slot, which is the type of connection usually used in walls where the separation is required only in a direction parallel to the wall plane. Details (b) and (c) show means for achieving the transfer of vertical gravity forces through the joint while permitting movement in a horizontal direction. Detail (d) shows a means for achieving a connection in one horizontal direction while permitting movement in the perpendicular direction.

8.3 Soil and Foundation Problems

The general problems of foundations for lateral resistive structures are discussed in Section 5.8 of Chapter 5. Following is a discussion of a few of the special problems that are occasionally encountered.

Figure 8.11 illustrates the typical problem of tying isolated

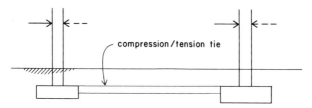

FIGURE 8.11 Tying of isolated footings.

footings, which is done primarily for the purpose of ensuring that the building structure move as a single mass with respect to the ground. It is also sometimes done in order to allow for the sharing of the lateral load when the load on a single footing cannot be resisted by that footing alone.

In many cases the existence of basement walls, grade walls, wall footings, or other parts of the building substructure make the addition of ties unnecessary. When individual footings are truly isolated, the tie member is designed as both a tension and a compression member. The required size of the concrete cross section and the main reinforcing is usually determined by the column action in compression. The tension tie is provided by extending the tie reinforcing into the footings in dowel action.

When calculated lateral loads exist for the footings, they may be used as the design loads on the ties. Section 2312(j)3.B of the *UBC* requires ties for pile caps and caissons to be designed for a minimum load of 0.10 of the vertical load on the heaviest loaded element. For isolated footings under structural elements that are not part of the lateral resistive structural system, there is no quantifiable basis for the tie design. Such footings are usually designed using the minimum requirements for concrete compressive members.

For foundations that do not use bearing footings, there are some special problems with regard to lateral loads. Figure 8.12 shows three such situations: drilled caissons, driven piles, and pole-type structures. Direct resistance to horizontal movement by the ground in all these cases can be developed only as passive lateral soil pressure. Where the upper soil strata are highly compressible, which is usually the case that causes piles and caissons

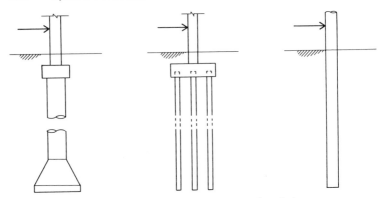

FIGURE 8.12   Deep, penetration-type foundations.

to be used, this resistance is quite limited. In addition, the construction of all three types of foundation shown results in considerable disturbance of the upper ground mass, which further lowers its effectiveness for passive resistance.

The tying of the pole structure is usually performed above the level of the ground, by the rest of the structure or by added ties and struts. The use of piles and caissons usually involves two considerations: the actual lateral load and the need for lateral stability under the vertical load. Tying is required for both of these reasons, and the previously cited *UBC* requirement is based on the latter problem—that of stability under the vertical load. Any calculated lateral loads delivered to the tops of the pile and caisson foundations should be carried away to other parts of the construction for development of the actual lateral resistance, which is essentially similar to the procedure used for isolated footings.

Another special soil problem occurs when there is a seismic load on retaining walls. As shown on the left in Figure 8.13, in designing for gravity loads, a lateral soil pressure is usually assumed on the basis of an equivalent horizontal fluid pressure. If a surcharge exists, or if the ground slopes to the wall as shown in the illustration, the level of the equivalent fluid is assumed to be somewhere above the true ground level at the back of the wall.

When lateral load has been caused by seismic movement, the

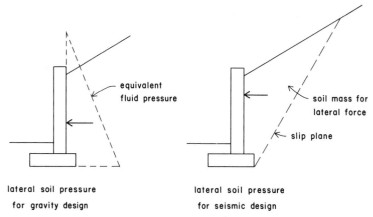

equivalent
fluid pressure

soil mass for
lateral force

slip plane

lateral soil pressure
for gravity design

lateral soil pressure
for seismic design

FIGURE 8.13 Lateral load conditions for retaining walls—gravity and seismic loads.

earth mass behind the wall has the potential of delivering a horizontal force while simultaneously offering gravity resistance to the overturn of the wall. The illustration on the right in Figure 8.13 shows the usual assumption for the soil failure mechanism that defines the potential mass whose weight develops the lateral seismic force. In a conservative design, for the analysis of the wall, this force should be added to the lateral seismic forces of the wall and footing weights.

Figure 8.14 illustrates the situation of a footing in a hillside

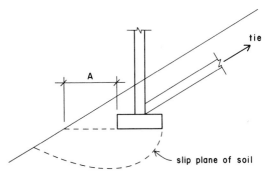

tie

A

slip plane of soil

FIGURE 8.14 Bearing footings in hillside conditions.

location that could be a wall footing or an isolated column footing. The dimension "A" in the figure, called the *daylight dimension,* must be sufficient to provide resistance to the failure of the soil as shown. Lateral load on the footing will further aggravate this type of failure if "A" is too small. The preferred solution, if the construction permits it, is to use a tension tie to transfer the lateral effect to some other part of the structure. Otherwise, the level of the footing should be lowered until a conservative distance is developed for the daylight dimension.

References

IIIIIIIIIIIIIIIIIIIIIIIIIIIIIIIIIIIIIIIIIIIIIIIIIIIIIIIIIIIIII

1. *Uniform Building Code,* 1979 ed., International Conference of Building Officials, 5360 South Workman Mill Road, Whittier, CA 90601.
2. Harry Parker, *Simplified Engineering for Architects and Builders,* 5th ed. (prepared by Harold D. Hauf), Wiley, New York, 1975.
3. Charles G. Ramsey and Harold R. Sleeper, *Architectural Graphic Standards,* 6th ed., Wiley, New York, 1970.
4. *Recommended Lateral Force Requirements and Commentary,* 1975 ed., Seismology Committee, Structural Engineers Association of California, 171 Second Street, San Francisco, CA 94105.
5. *Manual of Steel Construction,* 7th ed., American Institute of Steel Construction, 1221 Avenue of the Americas, New York, NY 10020.
6. *Building Code Requirements for Reinforced Concrete,* ACI 318–77, American Concrete Institute, P.O. Box 4754 Redford Station, Detroit, MI 48219.
7. *Concrete Masonry Design Manual,* 1974 ed., Concrete Masonry Association of California and Nevada, 83 Scripps Drive, Suite 303, Sacramento, CA 95825.
8. *Western Woods Use Book,* Western Woods Products Association, 1500 Yeon Building, Portland, OR 97204.
9. *Inryco Lateral Diaphragm Data Manual 20–2,* Inryco, Inc., Building Panels Division, P.O. Box 393, Milwaukee, WI, 53201.
10. Catalog No. 79H–1, *Structural Design and Load Values for Simpson Strong-Tie Connectors,* Simpson Company, P.O. Box 1568, San Leandro, CA 94577.
11. E. L. Houghton and N. B. Carruthers, *Wind Forces on Buildings and Structures,* Wiley, New York, 1976.
12. Robert Iacopi, *Earthquake Country,* 3rd ed., Lane Book Co., Menlo Park, CA, 1971.
13. Norman B. Green, *Earthquake Resistant Building Design and Construction,* Van Nostrand, New York, 1978.

265

14. David J. Dowrick, *Earthquake Resistant Design,* Wiley, New York, 1977.
15. Wolfgang Schueller, *High-Rise Building Structures,* Wiley, New York, 1977.
16. *Seismic Design for Buildings,* Army Technical Manual TM 5–809–12, Departments of the Army, Navy and Air Force, 1973.

Glossary

||||||||||||||||||||||||||||||||||||||||||||||||||||||||||||

The material presented here constitutes a brief dictionary of words and terms frequently encountered in discussions of the design of structures to withstand wind and earthquakes. Many of the words and terms have reasonably well-established meanings in the scientific and engineering literature. In those cases, we have tried to be consistent with the accepted usage. In some cases, however, words and terms are given somewhat different meanings by different authors, by different professional groups, in different fields of study, in different countries, and so on. In these situations, we have given the definitions as used by the authors of this work so that the reader may be clear as to our meaning.

In some cases, words or terms are commonly misused with regard to their precise meaning. We have generally used such words and terms as they are broadly understood, but we have given both the correct and popular definitions in some cases.

To be as clear as possible in its requirements, the *Uniform Building Code* occasionally gives its own definitions. Reference should be made to these definitions in interpreting *UBC* requirements.

For a fuller explanation of some of these words and terms, the reader should use the Index to find the related discussion in the text.

Acceleration.   The rate of change of the velocity, expressed as the first derivative of the velocity (dv/dt) or as the second derivative of the displacement (d^2s/dt^2). Acceleration of the ground surface is more significant than its displacement during an earthquake because it relates more directly to the force effect. $F = ma$, as a dynamic force.

Aerodynamic.   Refers to fluid flow effects of the air mass, similar to current effects in running water.

Anchorage.   Refers to attachment for resistance to movement; usually a result of uplift, overturn, sliding, or horizontal separation. Tie down, or hold down, refers to anchorage against uplift or overturn. Positive anchorage generally refers to direct fastening that does not easily loosen.

Aseismic.   The correct word for description of resistance to seismic effects. Building design actually consists of *aseismic* design, although the term seismic design is more commonly used.

Base shear.   See the *UBC* definition of *base* in Section 2312(b).

Battering.   Describes the effect that occurs when two elements in separate motion bump into each other repeatedly, such as two adjacent parts of a structure during an earthquake. Also called *pounding* or *hammering*.

Box system.   A structural system in which lateral loads are not resisted by a vertical load-bearing space frame but, rather, by shear walls or a braced frame.

Bracing.   In structural design usually refers to the resistance to movements caused by lateral forces or by the effects of buckling, torsional rotation, sliding, and so on.

Brittle fracture.   Sudden, ultimate failure in tension or shear. The basic structural behavior of so-called brittle materials.

Buffeting.   Refers to wind effects caused by turbulent air flow or by changes in the wind direction that result in whipping, rocking, and so on.

Collector. A force transfer element that functions to collect loads from a horizontal diaphragm and distribute them to the vertical elements of the lateral resistive system.

Continuity. Most often used to describe structures or parts of structures that have behavior characteristics influenced by the monolithic, continuous nature of adjacent elements, such as continuous, vertical, multistory columns, continuous, multispan beams, and rigid frames.

Damping. The effect that causes a decrease in the amplitude of succeeding cycles of harmonic motion. May be deliberately induced by devices such as shock absorbers.

Deflection. Generally refers to the lateral movement of a structure caused by loads, such as the vertical sag of a beam, the bowing of a surface under wind pressure, or the lateral sway of a rigid frame. The total horizontal deflection at the top of a structure caused by lateral loads is also called *drift*.

Degree of freedom. See *Freedom*.

Density. See *Mass*.

Displacement. Movement away from some fixed reference point. Motion is described mathematically as a displacement/time function. See *Velocity*, *Acceleration*.

Drag. Generally refers to wind effects on surfaces parallel to the wind direction. Ground drag refers to the effect of the ground surface in slowing the wind velocity near ground level.

Drag strut. A structural member used to transfer lateral load across the building and into some part of the vertical system. See also *Collector*.

Drift. See *Deflection*.

Diaphragm. A solid surface element (plywood deck, masonry wall, etc.) used to resist forces in its own plane by spanning or by cantilevering.

Ductile. Describes the load/strain behavior that results from the plastic yielding of materials or connections. To be significant, the plastic strain prior to failure should be considerably more than the elastic strain up to the point of plastic yield.

Dynamic. Usually used to characterize load effects or structural behaviors that are nonstatic in nature. That is, they involve time-related considerations, such as vibrations, energy effects versus simple force, and so on.

Earthquake. The common term used to describe subterranean ground shocks that result in movement of the ground surface. The shock is usually the result of a ground fault whose location is called the *epicenter* of the quake. The magnitude of the energy release at the epicenter is the basis for rating the size, or relative seriousness, of the quake. This energy release is measured on a scale such as the Richter scale.

Elastic. Used to describe two aspects of stress/strain behavior. The first is a constant stress/strain proportionality, or constant modulus of elasticity, as represented by a straight line form of the stress/strain graph. The second is the limit within which all of the strain is recoverable; that is, there is no permanent deformation. The latter phenomenon may occur even though the stress/strain relationship is nonlinear.

Energy. Capacity for doing work; what is used up when work is done. Occurs in various forms: mechanical, heat, chemical, electrical, and so on.

Epicenter. See *Earthquake*.

Equilibrium. A balanced state or condition, usually used to describe a situation in which opposed effects neutralize each other to produce a net effect of zero.

Equivalent static force analysis. The technique by which a dynamic effect is translated into a hypothetical (equivalent) static effect that produces a similar result.

Fault. The subterranean effect that produces an earthquake. Consists of a slippage, cracking, sudden strain release, and so on. See *Earthquake*.

Flexible. See *Stiffness*.

Flutter. Flapping, vibration type of movement of an object in high wind. Essentially a resonant behavior. See *Vibration*.

Force. An effort that tends to change the shape or the state of motion of an object.

Freedom. In structures, usually refers to the lack of some type of resistance or constraint. In static analysis, the connections between members and the supports of the structure are qualified as to type, or degree, of freedom. Thus, the terms *fixed support*, *pinned support*, and *sliding support* are used to qualify the types of movement resisted. In dynamic analysis the degree of freedom is an important factor in determining the dynamic response of a structure.

Frequency. In harmonic motion (bouncing springs, vibrating strings, swinging pendulums, etc.), the number of complete cycles of motion per unit of time. See *Vibration*.

Fundamental period. See *Period*.

Geophysical. Refers to the physical behavior characteristics of the ground surface and of subterranean masses.

Gust. An increase, or surge, of short duration, in the wind velocity.

Hammering. See *Battering*.

Header. Usually used to describe a horizontal element over an opening in a wall or at the edge of an opening in a roof or floor.

Hold down. See *Anchorage*.

Inelastic. See *Stress/strain behavior*.

Insurance. What to have a lot of, if you design or own structures.

Lateral. Literally means to the side or from the side. Often used in reference to something that is perpendicular to a major axis or direction. With reference to the vertical direction of the gravity forces, wind, earthquakes, and horizontally directed soil pressures are called lateral effects.

Let-in bracing. Diagonal boards nailed to studs to provide trussed bracing in the wall plane. In order not to interfere with the surfacing materials of the wall, they are usually notched in, or let in, to the stud faces.

Mass. The dynamic property of an object that causes it to resist changes in its state of motion. This resistance is called *inertia*. The magnitude of the mass per unit volume of the object is called its *density*. Dynamic force is defined by $F = ma$, or force equals mass times acceleration. Weight is defined as the force produced by the acceleration of gravity; thus, $W = mg$.

Natural period.   See *Period*.

Normal.   1. The ordinary, usual, unmodified state of something. 2. Perpendicular; such as pressure normal to a surface, stress normal to a cross section, and so on.

Occupancy.   In building code language refers to the use of a building as a residence, school, office, and so on.

Occupancy importance factor (I).   *UBC* term used in the basic equation for seismic force: $V = ZIKCSW$. Accounts for possible increased concern for certain occupancies.

Overturn.   The toppling, or tipping over, effect of lateral loads.

Parapet.   The extension of a wall plane or the roof edge facing above the roof level.

Period (of vibration.)   The total elapsed time for one full cycle of vibration. For an elastic structure in simple, single-mode vibration, the period is a constant (called the *natural* or *fundamental period*) and is independent of the magnitude of the amplitude, of the number of cycles, and of most damping or resonance effects. See *Vibration*.

Pier.   A short, stocky column. Section 2417(d) of the *UBC* defines a masonry wall as a pier if its plan length is less than three times the wall thickness. Otherwise, it would have to be designed as a column.

Positive anchorage.   See *Anchorage*.

Pounding.   See *Battering*.

Pressure.   A force distributed over, and normal to, a surface.

Relative stiffness.   See *Stiffness*.

Resonance.   See *Vibration*.

Response spectra.   See *Spectrum*.

Richter scale.   A log-based measuring system for evaluation of the relative energy level of an earthquake at its center of origin (epicenter).

Risk.   The degree of probability of loss due to some potential hazard. The risk of an earthquake in a particular geographic area is the basis for the *Z* factor in the *UBC* equation for seismic force: $V = ZIKCSW$.

Rotation.   Motion in a circular path. Also used to describe a twisting, or torsional, effect. See *Torsion*.

Seismic.   Pertaining to ground shock. See *Aseismic*.

Separation.   Often used in structural design to denote situations in which parts of a structure are made to act independently. Partial separation refers to a controlled separation that allows for some interactions but permits independence for other actions, such as a connection that transmits vertical forces but not horizontal ones. The separation may be dimensionally controlled to allow for a specific amount of movement.

Shear.   A force effect that is lateral (perpendicular) to the major axis of a structure, or one that involves a slipping effect, as opposed to a push–pull effect. Wind and earthquake forces are sometimes visualized as shear effects on a building because they are perpendicular to the major vertical (gravity) axis of the building.

Site response factor (S).   A *UBC* term used in the basic equation for seismic force: $V = ZIKCSW$. Accounts for the effect of the period of the ground mass under the building.

Space frame.   An ambiguous term, used variously to describe three-dimensional structures. The *UBC* uses a particular definition in Section 2312 in classifying structural systems.

Spectrum.   In seismic analysis, generally refers to the curve that describes the actual dynamic force effect on a structure as a function of variation in its fundamental period. Response spectra are the family of curves produced by various degrees of damping. This represents the basis for determining the *C* factor in the *UBC* equation for seismic force: $V = ZIKCSW$.

Stability.   Refers to the inherent capability of a structure to develop force resistance as a property of its form, orientation, articulation of its parts, type of connections, methods of support, and so on. Is not directly related to quantified strength or stiffness, except when the actions involve the buckling of elements of the structure.

Static.   The state that occurs when the velocity is zero; thus, no motion is occurring. Is generally used to refer to situations in which no change is occurring.

Stiffness.   In structures, refers to resistance to deformation, as opposed to strength, which refers to resistance to force. A lack of stiffness indicates a flexible structure. Relative stiffness usually refers to the comparative deformation of two or more structural elements that share a load.

Strain.   Deformation resulting from stress. Is usually measured as a percentage of deformation, called *unit strain* or *unit deformation*, and is thus dimensionless.

Stress.   The mechanism of force within the material of a structure. Is visualized as a pressure effect (if tension or compression) or a shear effect on the surface of a unit of the material and is quantified in terms of force per unit area.

Stress/strain behavior.   The relation of stress to strain in a material or a structure. Is usually visually represented by a stress/strain graph covering the range from no load to failure. Various aspects of the form of the graph define particular behavioral properties. A straight line indicates an elastic relationship; a curve indicates inelastic behavior. A sudden bend in the graph usually indicates a plastic strain, or yield, which results in some permanent deformation. The slope of the graph is defined as the modulus of elasticity of the material.

Tie down.   See *Anchorage*.

Torsion.   Moment effect involving twisting or rotation that is in a plane perpendicular to the major axis of an element. Lateral loads produce torsion on a building when they tend to twist it about its vertical axis. This occurs when the centroid of the load does not coincide with the center of stiffness of the vertical elements of the lateral load-resisting structural system.

Ultimate strength.   Usually used to refer to the maximum static force resistance of a structure at the time of failure. This limit is the basis for the so-called strength design methods, as compared to the stress design methods that use some established stress limit, called the design stress, working stress, permissible stress, and so on.

Uplift.   Usually refers to a net upward force effect caused when suction pressure from wind, or overturn effects from any lateral load, overcome the downward gravity effects.

Velocity.   The time rate of a motion, also commonly called *speed*.

Vibration.   The cyclic, rhythmic motion of a body such as a spring. Occurs when the body is displaced from some neutral position and seeks to restore itself to a state of equilibrium when released. In its pure form it occurs as a harmonic motion with a characteristic behavior described by the cosine form of the displacement/time graph of the motion. The magnitude of linear displacement from the neutral position is called the *amplitude*. The time elapsed for one full cycle of motion is called the *period*. The number of cycles occurring in one second is called the *frequency*. Effects that tend to reduce the amplitude of succeeding cycles are called *damping*. The increase of amplitude in successive cycles is called a *resonant effect*.

Yield.   See *Stress/strain behavior*.

Zone.   Usually refers to a bounded area on a surface, such as the ground surface or the plan of a level of a building.

Appendix

||||||||||||||||||||||||||||||||||||||||||||||||||||||||||||||||

This Appendix contains some of the materials that have been used as references in the design examples in Chapters 6 and 7. These materials are provided for the convenience of readers to whom the references may not be available. If they are available, however, it is highly recommended that information be taken directly from the references, most of which are frequently revised and contain additional explanations and examples of their use. Complete information for the publications from which these materials are taken is given in the References following Chapter 8.

The materials presented in this section are reprinted or adapted from the following sources:

Excerpts from the *Uniform Building Code,* 1979 edition, are reprinted with permission of the publishers, International Conference of Building Officials, 5360 So. Workman Mill Road, Whittier, CA 90601.

Table for lateral load design data for steel deck is reprinted from *Inryco Lateral Diaphragm Data Manual 20-2* with permission of the publishers, Inryco, Inc.

Tables of stiffness factors for masonry piers are reprinted from *Concrete Masonry Design Manual,* 1974 edition, with permission of the publishers, Concrete Masonry Association of California and Nevada.

Adaptation of data for various fastening devices for wood from the 1979 catalog of Simpson Strong-Tie Connectors is with the permission of the Simpson Company.

Anchorage of Concrete or Masonry Walls

Sec. 2310. Concrete or masonry walls shall be anchored to all floors and roofs which provide lateral support for the wall. Such anchorage shall provide a positive direct connection capable of resisting the horizontal forces specified in this chapter or a minimum force of 200 pounds per lineal foot of wall, whichever is greater. Walls shall be designed to resist bending between anchors where the anchor spacing exceeds 4 feet. Required anchors in masonry walls of hollow units or cavity walls shall be embedded in a reinforced grouted structural element of the wall. See Section 2312 (j) 2 D and 2312 (j) 3 A.

Wind Design

Sec. 2311. (a) General. Buildings or structures shall be designed to withstand the minimum horizontal and uplift pressures set forth in Table No. 23-F and this section allowing for wind from any direction. The wind pressures set forth in Table No. 23-F are minimum values and shall be adjusted by the building official for areas subjected to higher wind pressures. When the form factor, as determined by wind tunnel tests or other recognized methods, indicates vertical or horizontal loads of lesser or greater severity than those produced by the loads herein specified, the structure may be designed accordingly.

(b) **Horizontal Wind Pressure.** For purposes of design, the wind pressure shall be taken upon the gross area of the vertical projection of that portion of the building or structure measured above the average level of the adjoining ground.

(c) **Uplift Wind Pressure.** Roofs of all enclosed buildings or structures shall be designed and constructed to withstand pressures acting upward normal to the surface equal to three-fourths of the values set forth in Table No. 23-F for the height zone under consideration. An enclosed building shall be defined as a building enclosed at the perimeter with solid exterior walls. Openings are permitted in the solid exterior wall, provided they are glazed or protected with door assemblies.

Roofs of unenclosed buildings, roof overhangs, architectural projections, eaves, canopies, cornices, marquees or similar structures unenclosed on one or more sides shall be designed and constructed to withstand upward pressures equal to one and one-fourth times those values set forth in Table No. 23-F.

The upward pressures shall be assumed to act over the entire roof area.

(d) **Roofs with Slopes Greater than 30 Degrees.** Roofs or sections of roofs with slopes greater than 30 degrees shall be designed and constructed to withstand pressures, acting inward normal to the surface, equal to those specified for the height zone in which the roof is located, and applied to the windward slope only.

(e) **Anchorage Requirements.** Adequate anchorage of the roof to walls and columns, and of walls and columns to the foundations to resist overturning, uplift and sliding shall be provided in all cases.

(f) **Solid Towers.** Chimneys, tanks and solid towers shall be designed and constructed to withstand the pressures as specified by this section, multiplied by the factors set forth in Table No. 23-G.

(g) **Open Frame Towers.** Radio towers and other towers of trussed construction shall be designed and constructed to withstand wind pressures specified in this section, multiplied by the shape factors set forth in Table No. 23-H.

Wind pressures shall be applied to the total normal projected area of all the elements of one face (excluding ladders, conduits, lights, elevators, etc., which shall be accounted for separately by using the indicated factor for these individual members).

(h) **Miscellaneous Structures.** Fences less than 12 feet in height, greenhouses, lath houses and agricultural buildings shall be designed for the horizontal wind pressures as set forth in Table No. 23-F except that, if the height zone is 20 feet or less, two-thirds of the first line of listed values may be used. The structures shall be designed to withstand an uplift wind pressure equal to three-fourths of the horizontal pressure.

(i) **Moment of Stability.** The overturning moment calculated from the wind pressure shall in no case exceed two-thirds of the dead load resisting moment.

The weight of earth superimposed over footings may be used to calculate the dead load resisting moment.

(j) **Combined Wind and Live Loads.** For the purpose of determining stresses, all vertical design loads except the roof live load and crane loads shall be considered as acting simultaneously with the wind pressure.

> **EXCEPTION:** Where snow loading is required in the design of roofs, at least 50 percent of such snow load shall be considered acting in combination with the wind load. The building official may require that a greater percentage of snow load be considered due to local conditions.

Earthquake Regulations

Sec. 2312. (a) General. Every building or structure and every portion thereof shall be designed and constructed to resist stresses produced by lateral forces as provided in this section. Stresses shall be calculated as the effect of a force applied horizontally at each floor or roof level above the base. The force shall be assumed to come from any horizontal direction.

Structural concepts other than set forth in this section may be approved by the building official when evidence is submitted showing that equivalent ductility and energy absorption are provided.

Where prescribed wind loads produce higher stresses, such loads shall be used in lieu of the loads resulting from earthquake forces.

(b) **Definitions.** The following definitions apply only to the provisions of this section:

BASE is the level at which the earthquake motions are considered to be imparted to the structure or the level at which the structure as a dynamic vibrator is supported.

BOX SYSTEM is a structural system without a complete vertical load-carrying space frame. In this system the required lateral forces are resisted by shear walls or braced frames as hereinafter defined.

BRACED FRAME is a truss system or its equivalent which is provided to resist lateral forces in the frame system and in which the members are subjected primarily to axial stresses.

DUCTILE MOMENT-RESISTING SPACE FRAME is a moment-resisting space frame complying with the requirements for a ductile moment-resisting space frame as given in Section 2312 (j).

ESSENTIAL FACILITIES—See Section 2312 (k).

LATERAL FORCE-RESISTING SYSTEM is that part of the structural system assigned to resist the lateral forces prescribed in Section 2312 (d) 1.

MOMENT-RESISTING SPACE FRAME is a vertical load-carrying space frame in which the members and joints are capable of resisting forces primarily by flexure.

SHEAR WALL is a wall designed to resist lateral forces parallel to the wall.

SPACE FRAME is a three-dimensional structural system without bearing walls, composed of interconnected members laterally supported so as to function as a complete self-contained unit with or without the aid of horizontal diaphragms or floor-bracing systems.

VERTICAL LOAD-CARRYING SPACE FRAME is a space frame designed to carry all vertical loads.

(c) **Symbols and Notations.** The following symbols and notations apply only to the provisions of this section:

C = Numerical coefficient as specified in Section 2312 (d) 1.

C_p = Numerical coefficient as specified in Section 2312 (g) and as set forth in Table No. 23-J.

D = The dimension of the structure, in feet, in a direction parallel to the applied forces.

δ_i = Deflection at level i relative to the base, due to applied lateral forces, Σf_i, for use in Formula (12-3).

$F_i F_n F_x$ = Lateral force applied to level i, n or x, respectively.

F_p = Lateral forces on a part of the structure and in the direction under consideration.

F_t = That portion of V considered concentrated at the top of the structure in addition to F_n.

f_i = Distributed portion of a total lateral force at level i for use in Formula (12-3).

g = Acceleration due to gravity.

$h_i h_n h_x$ = Height in feet above the base to level i, n or x respectively.

I = Occupancy Importance Factor as set forth in Table No. 23-K.

K = Numerical coefficient as set forth in Table No. 23-I.

Level i

l = Level of the structure referred to by the subscript i.

i = 1 designates the first level above the base.

Level n

= That level which is uppermost in the main portion of the structure.

Level x

= That level which is under design consideration.

x = 1 designates the first level above the base.

N = The total number of stories above the base to level n.

S = Numerical coefficient for site-structure resonance.

T = Fundamental elastic period of vibration of the building or structure in seconds in the direction under consideration.

T_s = Characteristic site period.

V = The total lateral force or shear at the base.

W = The total dead load as defined in Section 2302 including the partition loading specified in Section 2304 (d) where applicable.

EXCEPTION: W shall be equal to the total dead load plus 25 percent of the floor live load in storage and warehouse occupancies. Where the design snow load is 30 psf or less, no part need be included in the value of W. Where the snow load is greater than 30 psf, the snow load shall be included; however, where the snow load duration warrants, the building official may allow the snow load to be reduced up to 75 percent.

w_iw_x = That portion of W which is located at or is assigned to level i or x respectively.

W_p = The weight of a portion of a structure or nonstructural component.

Z = Numerical coefficient dependent upon the zone as determined by Figures No. 1, No. 2 and No. 3 in this chapter. For locations in Zone No. 1, $Z = \frac{3}{16}$. For locations in Zone No. 2, $Z = \frac{3}{8}$. For locations in Zone No. 3, $Z = \frac{3}{4}$. For locations in Zone No. 4, $Z = 1$.

(d) Minimum Earthquake Forces for Structures. Except as provided in Section 2312 (g) and (i), every structure shall be designed and constructed to resist minimum total lateral seismic forces assumed to act nonconcurrently in the direction of each of the main axes of the structure in accordance with the following formula:

$$V = ZIKCSW \dots\dots\dots\dots (12\text{-}1)$$

The value of K shall be not less than that set forth in Table No. 23-I. The value of C and S are as indicated hereafter except that the product of CS need not exceed 0.14.

The value of C shall be determined in accordance with the following formula:

$$C = \frac{1}{15\ \sqrt{T}} \qquad \dots \dots \dots \dots \dots \text{(12-2)}$$

The value of C need not exceed 0.12.

The period T shall be established using the structural properties and deformational characteristics of the resisting elements in a properly substantiated analysis such as the following formula:

$$T = 2\pi \sqrt{\left(\sum_{i=1}^{n} \omega_i \delta_i^2\right) \div \left(g \sum_{i=1}^{n} f_i \delta_i\right)}$$

$$\dots \dots \dots \dots \dots \dots \dots \dots \dots \text{(12-3)}$$

where the values of f_i represent any lateral force distributed approximately in accordance with the principles of Formulas (12-5), (12-6) and (12-7) or any other rational distribution. The elastic deflections, δ_i, shall be calculated using the applied lateral forces, f_i.

In the absence of a determination as indicated above, the value of T for buildings may be determined by the following formula:

$$T = \frac{0.05 h_n}{\sqrt{D}} \qquad \dots \dots \dots \dots \dots \text{(12-3A)}$$

Or in buildings in which the lateral force-resisting system consists of ductile moment-resisting space frames capable of resisting 100 percent of the required lateral forces and such system is not enclosed by or adjoined by more rigid elements tending to prevent the frame from resisting lateral forces:

$$T = 0.10N \dots \dots \dots \dots \dots \text{(12-3B)}$$

The value of S shall be determined by the following formulas, but shall be not less than 1.0:

For $T/T_s = 1.0$ or less $\quad S = 1.0 + \dfrac{T}{T_s} - 0.5 \left[\dfrac{T}{T_s}\right]^2$

$$\dots \dots \dots \dots \dots \dots \text{(12-4)}$$

For T/T_s greater than 1.0 $\quad S = 1.2 + 0.6 \dfrac{T}{T_s} - 0.3 \left[\dfrac{T}{T_s}\right]^2$

$$\dots \dots \dots \dots \dots \dots \text{(12-4A)}$$

WHERE:

T in Formulas (12-4) and (12-4A) shall be established by a properly substantiated analysis but T shall be not less than 0.3 second.

The range of values of T_s may be established from properly substantiated geotechnical data, in accordance with U.B.C. Standard No. 23-1, except that T_s shall not be taken as less than 0.5 second nor more than 2.5 seconds. T_s shall be that value within the range of site periods, as determined above, that is nearest to T.

When T_s is not properly established, the value of S shall be 1.5.

> **EXCEPTION:** Where T has been established by a properly substantiated analysis and exceeds 2.5 seconds, the value of S may be determined by assuming a value of 2.5 seconds for T_s.

(e) **Distribution of Lateral Forces. 1. Structures having regular shapes or framing systems.** The total lateral force V shall be distributed over the height of the structure in accordance with Formulas (12-5), (12-6) and (12-7).

$$V = F_t + \sum_{i=1}^{n} F_i \dots\dots\dots\dots\dots (12\text{-}5)$$

The concentrated force at the top shall be determined according to the following formula:

$$F_t = 0.07TV \dots\dots\dots\dots\dots (12\text{-}6)$$

F_t need not exceed $0.25V$ and may be considered as 0 where T is 0.7 second or less. The remaining portion of the total base shear V shall be distributed over the height of the structure including level n according to the following formula:

$$F_x = \frac{(V - F_t)\, w_x h_x}{\sum_{i=1}^{n} w_i h_i} \dots\dots\dots\dots\dots (12\text{-}7)$$

At each level designated as x, the force F_x shall be applied over the area of the building in accordance with the mass distribution on that level.

2. Setbacks. Buildings having setbacks wherein the plan dimension of the tower in each direction is at least 75 percent of the corresponding plan dimension of the lower part may be considered as uniform buildings without setbacks, provided other irregularities as defined in this section do not exist.

3. Structures having irregular shapes or framing systems. The distribution of the lateral forces in structures which have highly irregular shapes, large differences in lateral resistance or stiffness between adjacent stories, or other unusual structural features, shall be determined considering the dynamic characteristics of the structure.

4. Distribution of horizontal shear. Total shear in any horizontal plane shall be distributed to the various elements of the lateral force-resisting system in proportion to their rigidities considering the rigidity of the horizontal bracing system or diaphragm.

Rigid elements that are assumed not to be part of the lateral force-resisting system may be incorporated into buildings provided that their effect on the action of the system is considered and provided for in the design.

5. Horizontal torsional moments. Provisions shall be made for the increase in shear resulting from the horizontal torsion due to an eccentricity between the center of mass and the center of rigidity. Negative torsional shears shall be neglected. Where the vertical resisting elements depend on diaphragm action for shear distribution at any level, the shear-resisting elements shall be capable of resisting a torsional moment assumed to be equivalent to the story shear acting with an eccentricity of not less than 5 percent of the maximum building dimension at that level.

(f) Overturning. Every building or structure shall be designed to resist the overturning effects caused by the wind forces and related requirements specified in Section 2311 or the earthquake forces specified in this section, whichever governs.

At any level the incremental changes of the design overturning moment, in the story under consideration, shall be distributed to the various resisting elements in the same proportion as the distribution of the shears in the resisting system. Where other vertical members are provided which are capable of partially resisting the overturning moments, a redistribution may be made to these members if framing members of sufficient strength and stiffness to transmit the required loads are provided.

Where a vertical resisting element is discontinuous, the overturning moment carried by the lowest story of that element shall be carried down as loads to the foundation.

(g) Lateral Force on Elements of Structures and Nonstructural Components. Parts or portions of structures, nonstructural components and their anchorage to the main structural system shall be designed for lateral forces in accordance with the following formula:

$$F_p = ZIC_pW_p \dots \dots \dots \dots \dots (12-8)$$

The values of C_p are set forth in Table No. 23-J. The value of the I coefficient shall be the value used for the building.

> **EXCEPTIONS:** 1. The value of I for panel connectors shall be as given in Section 2312 (j) 3 C.
>
> 2. The value of I for anchorage of machinery and equipment required for life safety systems shall be 1.5.

The distribution of these forces shall be according to the gravity loads pertaining thereto.

For applicable forces on diaphragms and connections for exterior panels, refer to Sections 2312 (j) 2 D and 2312 (j) 3 C.

(h) **Drift and Building Separations.** Lateral deflections or drift of a story relative to its adjacent stories shall not exceed 0.005 times the story height unless it can be demonstrated that greater drift can be tolerated. The displacement calculated from the application of the required lateral forces shall be multiplied by $(1.0/K)$ to obtain the drift. The ratio $(1.0/K)$ shall be not less than 1.0.

All portions of structures shall be designed and constructed to act as an integral unit in resisting horizontal forces unless separated structurally by a distance sufficient to avoid contact under deflection from seismic action or wind forces.

(i) **Alternate Determination and Distribution of Seismic Forces.** Nothing in Section 2312 shall be deemed to prohibit the submission of properly substantiated technical data for establishing the lateral forces and distribution by dynamic analyses. In such analyses the dynamic characteristics of the structure must be considered.

(j) **Structural Systems.** 1. **Ductility requirements.** A. All buildings designed with a horizontal force factor $K = 0.67$ or 0.80 shall have ductile moment-resisting space frames.

B. Buildings more than 160 feet in height shall have ductile moment-resisting space frames capable of resisting not less than 25 percent of the required seismic forces for the structure as a whole.

> **EXCEPTION:** Buildings more than 160 feet in height in Seismic Zones Nos. 1 and 2 may have concrete shear walls designed in accordance with Section 2627 or braced frames designed in conformance with Section 2312 (j) 1 G of this code in lieu of a ductile moment-resisting space frame, provided a K value of 1.00 or 1.33 is utilized in the design.

C. In Seismic Zones No. 2, No. 3 and No. 4 all concrete space frames required by design to be part of the lateral force-resisting system and all concrete frames located in the perimeter line of vertical support shall be ductile moment-resisting space frames.

> **EXCEPTION:** Frames in the perimeter line of the vertical support of buildings designed with shear walls taking 100 percent of the design lateral forces need only conform with Section 2312 (j) 1 D.

D. In Seismic Zones No. 2, No. 3 and No. 4 all framing elements not required by design to be part of the lateral force-resisting system shall be investigated and shown to be adequate for vertical load-carrying capacity and induced moment due to $3/K$ times the distortions resulting from the code-required lateral forces. The rigidity of other elements shall be considered in accordance with Section 2312 (e) 4.

E. Moment-resisting space frames and ductile moment-resisting space frames may be enclosed by or adjoined by more rigid elements which would tend to prevent the space frame from resisting lateral forces where it can be shown that the action or failure of the more rigid elements will not impair the vertical and lateral load resisting ability of the space frame.

F. Necessary ductility for a ductile moment-resisting space frame shall be provided by a frame of structural steel with moment-resisting connections (complying with Section 2722 for buildings in Seismic Zones No. 3 and No. 4 or Section 2723 for buildings in Seismic Zones No. 1 and No. 2) or by a reinforced concrete frame (complying with Section 2626 for buildings in Seismic Zones No. 3 and No. 4 or Section 2625 for buildings in Seismic Zones No. 1 and No. 2).

> **EXCEPTION:** Buildings with ductile moment-resisting space frames in Seismic Zones No. 1 and No. 2 having an importance factor I greater than 1.0 shall comply with Section 2626 or 2722.

G. In Seismic Zones No. 3 and No. 4 and for buildings having an importance factor I greater than 1.0 located in Seismic Zone No. 2, all members in braced frames shall be designed for 1.25 times the force determined in accordance with Section 2312 (d). Connections shall be designed to develop the full capacity of the members or shall be based on the above forces without the one-third increase usually permitted for stresses resulting from earthquake forces.

Braced frames in buildings shall be composed of axially loaded bracing members of A36, A440, A441, A501, A572 (except Grades 60 and 65) or A588 structural steel; or reinforced concrete members conforming to the requirements of Section 2627.

H. Reinforced concrete shear walls for all buildings shall conform to the requirements of Section 2627.

I. In structures where $K = 0.67$ and $K = 0.80$, the special ductility requirements for structural steel or reinforced concrete specified in Section 2312 (j) 1 F, shall apply to all structural elements below the base which are required to transmit to the foundation the forces resulting from lateral loads.

2. Design requirements. A. Minor alterations. Minor structural alterations may be made in existing buildings and other structures, but the resistance to lateral forces shall be not less than that before such alterations were made, unless the building as altered meets the requirements of this section.

B. Reinforced masonry or concrete. All elements within structures located in Seismic Zones No. 2, No. 3 and No. 4 which are of masonry or concrete shall be reinforced so as to qualify as reinforced masonry or concrete under the provisions of Chapters 24 and 26. Principal reinforcement in masonry shall be spaced 2 feet maximum on center in buildings using a moment-resisting space frame.

C. Combined vertical and horizontal forces. In computing the effect of seismic force in combination with vertical loads, gravity load stresses induced in members by dead load plus design live load, except roof live load, shall be considered. Consideration should also be given to minimum gravity loads acting in combination with lateral forces.

D. **Diaphragms.** Floor and roof diaphragms and collectors shall be designed to resist the forces determined in accordance with the following formula:

$$F_{px} = \frac{\displaystyle\sum_{l=x}^{n} F_l}{\displaystyle\sum_{l=x}^{n} w_l} w_{px} \quad \ldots\ldots\ldots\ldots\ldots\ldots \text{(12-9)}$$

WHERE:

F_l = the lateral force applied to level l.

w_l = the portion of W at level l.

w_{px} = the weight of the diaphragm and the elements tributary thereto at level x, including 25 percent of the floor live load in storage and warehouse occupancies.

The force F_{px} determined from Formula (12-9) need not exceed $0.30ZIw_{px}$.

When the diaphragm is required to transfer lateral forces from the vertical resisting elements above the diaphragm to other vertical resisting elements below the diaphragm due to offsets in the placement of the elements or to changes in stiffness in the vertical elements, these forces shall be added to those determined from Formula (12-9).

However, in no case shall lateral force on the diaphragm be less than $0.14ZIw_{px}$.

Diaphragms supporting concrete or masonry walls shall have continuous ties between diaphragm chords to distribute, into the diaphragm, the anchorage forces specified in this chapter. Added chords may be used to form sub-diaphragms to transmit the anchorage forces to the main cross ties. Diaphragm deformations shall be considered in the design of the supported walls. See Section 2312 (j) 3 A for special anchorage requirements of wood diaphragms.

3. **Special requirements. A. Wood diaphragms providing lateral support for concrete or masonry walls.** Where wood diaphragms are used to laterally support concrete or masonry walls the anchorage shall conform to Section 2310. In Zones No. 2, No. 3 and No. 4 anchorage shall not be accomplished by use of toenails or nails subjected to withdrawal; nor shall wood framing be used in cross-grain bending or cross-grain tension.

B. **Pile caps and caissons.** Individual pile caps and caissons of every building or structure shall be interconnected by ties, each of which can carry by tension and compression a minimum horizontal force equal to 10 percent of the larger pile cap or caisson loading, unless it can be demonstrated that equivalent restraint can be provided by other approved methods.

C. **Exterior elements.** Precast or prefabricated nonbearing, nonshear wall panels or similar elements which are attached to or enclose the exterior shall be designed to resist the forces determined from Formula (12-8) and shall accommodate movements of the structure resulting from lateral forces or temperature changes. The concrete panels or other similar elements shall be supported by means of cast-in-place concrete or mechanical connections and fasteners in accordance with the following provisions:

Connections and panel joints shall allow for a relative movement between stories of not less than two times story drift caused by wind or $(3.0/K)$ times the calculated elastic story displacement caused by required seismic forces, or ½ inch, whichever is greater. Connections to permit movement in the plane of the panel for story drift shall be properly designed sliding connections using slotted or oversized holes or may be connections which permit movement by bending of steel or other connections providing equivalent sliding and ductility capacity.

Bodies of connectors shall have sufficient ductility and rotation capacity so as to preclude fracture of the concrete or brittle failures at or near welds.

The body of the connector shall be designed for one and one-third times the force determined by Formula (12-8). Fasteners attaching the connector to the panel or the structure such as bolts, inserts, welds, dowels, etc., shall be designed to insure ductile behavior of the connector or shall be designed for four times the load determined from Formula (12-8).

Fasteners embedded in concrete shall be attached to or hooked around reinforcing steel or otherwise terminated so as to effectively transfer forces to the reinforcing steel.

The value of the coefficient I shall be 1.0 for the entire connector assembly in Formula (12-8).

(k) **Essential Facilities.** Essential facilities are those structures or buildings which must be safe and usable for emergency purposes after an earthquake in order to preserve the health and safety of the general public. Such facilities shall include but not be limited to:

1. Hospitals and other medical facilities having surgery or emergency treatment areas.
2. Fire and police stations.
3. Municipal government disaster operation and communication centers deemed to be vital in emergencies.

The design and detailing of equipment which must remain in place and be functional following a major earthquake shall be based upon the requirements of Section 2312 (g) and Table No. 23-J. In addition, their design and detailing shall consider effects induced by structure drifts of not less than $(2.0/K)$ times the story drift caused by required seismic forces nor less than the story drift caused by wind. Special consideration shall also be given to relative movements at separation joints.

(l) **Earthquake-recording Instrumentations.** For earthquake recording instrumentations see Appendix, Section 2312 (l).

TABLE NO. 23-F — WIND PRESSURES FOR VARIOUS HEIGHT ZONES ABOVE GROUND[1]

HEIGHT ZONES (in feet)	WIND-PRESSURE MAP AREAS (pounds per square foot)						
	20	25	30	35	40	45	50
Less than 30	15	20	25	25	30	35	40
30 to 49	20	25	30	35	40	45	50
50 to 99	25	30	40	45	50	55	60
100 to 499	30	40	45	55	60	70	75
500 to 1199	35	45	55	60	70	80	90
1200 and over	40	50	60	70	80	90	100

[1]See Figure No. 4. Wind pressure column in the table should be selected which is headed by a value corresponding to the minimum permissible, resultant wind pressure indicated for the particular locality.

The figures given are recommended as minimum. These requirements do not provide for tornadoes.

TABLE NO. 23-G — MULTIPLYING FACTORS FOR WIND PRESSURES — CHIMNEYS, TANKS AND SOLID TOWERS

HORIZONTAL CROSS SECTION	FACTOR
Square or rectangular	1.00
Hexagonal or octagonal	0.80
Round or elliptical	0.60

TABLE NO. 23-H — SHAPE FACTORS FOR RADIO TOWERS AND TRUSSED TOWERS

TYPE OF EXPOSURE	FACTOR
1. Wind normal to one face of tower	
Four-cornered, flat or angular sections, steel or wood	2.20
Three-cornered, flat or angular sections, steel or wood	2.00
2. Wind on corner, four-cornered tower, flat or angular sections	2.40
3. Wind parallel to one face of three-cornered tower, flat or angular sections	1.50
4. Factors for towers with cylindrical elements are approximately two-thirds of those for similar towers with flat or angular sections	
5. Wind on individual members	
Cylindrical members	
Two inches or less in diameter	1.00
Over two inches in diameter	0.80
Flat or angular sections	1.30

289

TABLE NO. 23-I—HORIZONTAL FORCE FACTOR *K* FOR BUILDINGS OR OTHER STRUCTURES[1]

TYPE OR ARRANGEMENT OF RESISTING ELEMENTS	VALUE[2] OF *K*
1. All building framing systems except as hereinafter classified	1.00
2. Buildings with a box system as specified in Section 2312 (b)	1.33
3. Buildings with a dual bracing system consisting of a ductile moment-resisting space frame and shear walls or braced frames using the following design criteria: a. The frames and shear walls shall resist the total lateral force in accordance with their relative rigidities considering the interaction of the shear walls and frames b. The shear walls acting independently of the ductile moment-resisting portions of the space frame shall resist the total required lateral forces c. The ductile moment-resisting space frame shall have the capacity to resist not less than 25 percent of the required lateral force	0.80
4. Buildings with a ductile moment-resisting space frame designed in accordance with the following criteria: The ductile moment-resisting space frame shall have the capacity to resist the total required lateral force	0.67
5. Elevated tanks plus full contents, on four or more cross-braced legs and not supported by a building.	2.5[3]
6. Structures other than buildings and other than those set forth in Table No. 23-J	2.00

[1]Where wind load as specified in Section 2311 would produce higher stresses, this load shall be used in lieu of the loads resulting from earthquake forces.

[2]See Figures Nos. 1, 2 and 3 in this chapter and definition of *Z* as specified in Section 2312 (c).

[3]The minimum value of *KC* shall be 0.12 and the maximum value of *KC* need not exceed 0.25.

The tower shall be designed for an accidental torsion of 5 percent as specified in Section 2312 (e) 5. Elevated tanks which are supported by buildings or do not conform to type or arrangement of supporting elements as described above shall be designed in accordance with Section 2312 (g) using $C_p = .3$.

TABLE NO. 23-J—HORIZONTAL FORCE FACTOR C_p FOR ELEMENTS OF STRUCTURES AND NONSTRUCTURAL COMPONENTS

PART OR PORTION OF BUILDINGS	DIRECTION OF HORIZONTAL FORCE	VALUE OF C_p [1]
1. Exterior bearing and nonbearing walls, interior bearing walls and partitions, interior nonbearing walls and partitions —see also Section 2312 (j) 3 C. Masonry or concrete fences over 6 feet high	Normal to flat surface	0.3 [6]
2. Cantilever elements: a. Parapets	Normal to flat surfaces	0.8
b. Chimneys or stacks	Any direction	0.3
3. Exterior and interior ornamentations and appendages	Any direction	0.8
4. When connected to, part of, or housed within a building: a. Penthouses, anchorage and supports for chimneys and stacks and tanks, including contents	Any direction	0.3 [2] [3]
b. Storage racks with upper storage level at more than 8 feet in height, plus contents		
c. All equipment or machinery		
5. Suspended ceiling framing systems (applies to Seismic Zones Nos. 2, 3 and 4 only)	Any direction	0.3 [4]
6. Connections for prefabricated structural elements other than walls, with force applied at center of gravity of assembly	Any direction	0.3 [5]

[1] C_p for elements laterally self-supported only at the ground level may be two-thirds of value shown.

[2] W_p for storage racks shall be the weight of the racks plus contents. The value of C_p for racks over two storage support levels in height shall be 0.24 for the levels below the top two levels. In lieu of the tabulated values steel storage racks may be designed in accordance with U.B.C. Standard No. 27-11.

Where a number of storage rack units are interconnected so that there are a minimum of four vertical elements in each direction on each column line designed to resist horizontal forces, the design coefficients may be as for a building with K values from Table No. 23-1, $CS = 0.2$ for use in the formula $V = ZIKCSW$ and W equal to the total dead load plus 50 percent of the rack-rated capacity. Where the design and rack configurations are in accordance with this paragraph, the design provisions in U.B.C. Standard No. 27-11 do not apply.

[3] For flexible and flexibly mounted equipment and machinery, the appropriate values of C_p shall be determined with consideration given to both the dynamic properties of the equipment and

(Continued)

FOOTNOTES FOR TABLE 23-J—(Continued)

machinery and to the building or structure in which it is placed but shall be not less than the listed values. The design of the equipment and machinery and their anchorage is an integral part of the design and specification of such equipment and machinery.

For essential facilities and life safety systems, the design and detailing of equipment which must remain in place and be functional following a major earthquake shall consider drifts in accordance with Section 2312 (k).

⁴Ceiling weight shall include all light fixtures and other equipment which is laterally supported by the ceiling. For purposes of determining the lateral force, a ceiling weight of not less than 4 pounds per square foot shall be used.

⁵The force shall be resisted by positive anchorage and not by friction.

⁶See also Section 2309 (b) for minimum load and deflection criteria for interior partitions.

TABLE NO. 23-K
VALUES FOR OCCUPANCY IMPORTANCE FACTOR I

TYPE OF OCCUPANCY	I
Essential Facilities¹	1.5
Any building where the primary occupancy is for assembly use for more than 300 persons (in one room)	1.25
All others	1.0

See Section 2312 (k) for definition and additional requirements for essential facilities.

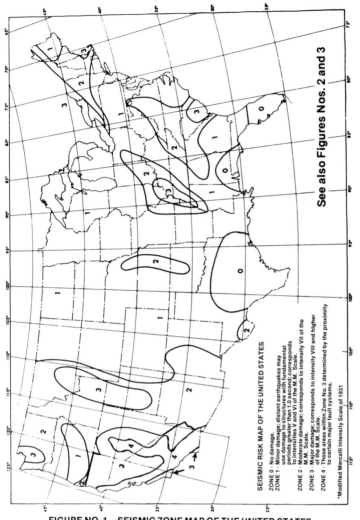

SEISMIC RISK MAP OF THE UNITED STATES

ZONE 0 - No damage.
ZONE 1 - Minor damage; distant earthquakes may
use damage to structures with fundamental
periods greater than 1.0 second; corresponds
to intensities V and VI of the M.M. Scale.

ZONE 2 - Moderate damage; corresponds to intensity VII of the
M.M. Scale.

ZONE 3 - Major damage; corresponds to intensity VIII and higher
of the M.M. Scale.

ZONE 4 - Those areas within Zone No. 3 determined by the proximity
to certain major fault systems.

*Modified Mercalli Intensity Scale of 1931

See also Figures Nos. 2 and 3

FIGURE NO. 1—SEISMIC ZONE MAP OF THE UNITED STATES
For areas outside of the United States, see Appendix Chapter 23

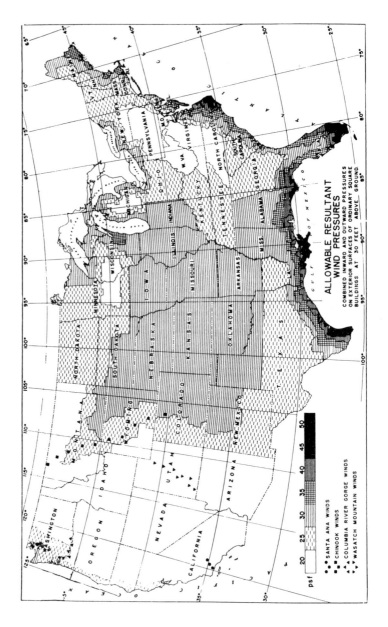

ALLOWABLE RESULTANT WIND PRESSURES

COMBINED INWARD AND OUTWARD PRESSURES ON EXTERIOR SURFACES OF ORDINARY SQUARE BUILDINGS AT 30 FEET ABOVE GROUND.

psf 20 25 30 35 40 45 50

●●● SANTA ANA WINDS
■■■ CHINOOK WINDS
▲▲▲ COLUMBIA RIVER GORGE WINDS
▼▼▼ WASATCH MOUNTAIN WINDS

	SPECIAL INSPECTION REQUIRED	
TYPE OF STRESS	Yes	No
1. Compression—Axial, Walls	See Section 2418	One-half of the values permitted under Section 2418
2. Compression—Axial, Columns	See Section 2418	One-half of the values permitted under Section 2418
3. Compression—Flexural	$0.33 f'_m$ but not to exceed 900	$0.166 f'_m$ but not to exceed 450
4. Shear: a. No shear reinforcement, Flexural[2]	$1.1\sqrt{f'_m}$ 50 Max.	25
Shear walls[3] $M/Vd \geq 1$[4]	$.9\sqrt{f'_m}$ 34 Max.	17
$M/Vd = 0$[4]	$2.0\sqrt{f'_m}$ 50 Max.	25
b. Reinforcing taking all shear, Flexural	$3.0\sqrt{f'_m}$ 150 Max.	75
Shear walls[3] $M/Vd \geq 1$[4]	$1.5\sqrt{f'_m}$ 75 Max.	35
$M/Vd = 0$[4]	$2.0\sqrt{f'_m}$ 120 Max.	60
5. Modulus of Elasticity[5]	$1000 f'_m$ but not to exceed 3,000,000	$500 f'_m$ but not to exceed 1,500,000
6. Modulus of Rigidity[5]	$400 f'_m$ but not to exceed 1,200,000	$200 f'_m$ but not to exceed 600,000
7. Bearing on full Area[6]	$0.25 f'_m$ but not to exceed 900	$0.125 f'_m$ but not to exceed 450
8. Bearing on ⅓ or less of area[6]	$0.30 f'_m$ but not to exceed 1200	$0.15 f'_m$ but not to exceed 600
9. Bond—Plain bars	60	30
10. Bond—Deformed	140	100

[1]Stresses for hollow unit masonry are based on net section.

[2]Web reinforcement shall be provided to carry the entire shear in excess of 20 pounds per square inch whenever there is required negative reinforcement and for a distance of one-sixteenth the clear span beyond the point of inflection.

[3]When calculating shear or diagonal tension stresses, shear walls which resist seismic forces shall be designed to resist 1.5 times the forces required by Section 2312 (d) 1.

(Continued)

FOOTNOTES FOR TABLE NO. 24-H—(Continued)

⁴M is the maximum bending moment occurring simultaneously with the shear load V at the section under consideration. Interpolate by straight line for M/Vd values between 0 and 1.

⁵Where determinations involve rigidity considerations in combination with other materials or where deflections are involved, the moduli of elasticity and rigidity under columns entitled "yes" for special inspection shall be used.

⁶This increase shall be permitted only when the least distance between the edges of the loaded and unloaded areas is a minimum of one-fourth of the parallel side dimension of the loaded area. The allowable bearing stress on a reasonably concentric area greater than one-third, but less than the full area, shall be interpolated between the values given.

Wood Diaphragms

Sec. 2514. (a) General. Lumber and plywood diaphragms may be used to resist horizontal forces in horizontal and vertical distributing or resisting elements, provided the deflection in the plane of the diaphragm, as determined by calculations, tests or analogies drawn therefrom, does not exceed the permissible deflection of attached distributing or resisting elements. See U.B.C. Standard No. 25-9 for a method of calculating the deflection of a blocked plywood diaphragm.

Permissible deflection shall be that deflection up to which the diaphragm and any attached distributing or resisting element will maintain its structural integrity under assumed load conditions, i.e., continue to support assumed loads without danger to occupants of the structure.

Connections and anchorages capable of resisting the design forces shall be provided between the diaphragms and the resisting elements. Openings in diaphragms which materially affect their strength shall be fully detailed on the plans and shall have their edges adequately reinforced to transfer all shearing stresses.

Size and shape of diaphragms shall be limited as set forth in Table No. 25-I.

In buildings of wood frame construction where rotation is provided for, the depth of the diaphragm normal to the open side shall not exceed 25 feet nor two-thirds the diaphragm width, whichever is the smaller depth. Straight sheathing shall not be permitted to resist shears in diaphragms acting in rotation.

> **EXCEPTIONS:** 1. One-story, wood-framed structures with the depth normal to the open side not greater than 25 feet may have a depth equal to the width.
>
> 2. Where calculations show that diaphragm deflections can be tolerated, the depth normal to the open end may be increased to a depth-to-width ratio not greater than 1½:1 for diagonal sheathing or 2:1 for special diagonal sheathed or plywood diaphragms.

In masonry or concrete buildings, lumber and plywood diaphragms shall not be considered as transmitting lateral forces by rotation.

Diaphragm sheathing nails or other approved sheathing connectors shall be driven flush but shall not fracture the surface of the sheathing.

(b) **Diagonally Sheathed Diaphragms.** 1. **Conventional construction.** Such lumber diaphragms shall be made up of 1-inch nominal sheathing boards laid at an angle of approximately 45 degrees to supports. Sheathing boards shall be directly nailed to each intermediate bearing member with not less than two 8d nails for 1-inch by 6-inch nominal boards and three 8d nails for boards 8 inches or wider; and in addition three 8d nails and four 8d nails shall be used for 6-inch and 8-inch boards, respectively, at the diaphragm boundaries. End joints in adjacent boards shall be separated by at least one joist or stud space, and there shall be at least two boards between joints on the same support. Boundary members at edges of diaphragms shall be designed to resist direct tensile or compressive chord stresses and shall be adequately tied together at corners.

Conventional lumber diaphragms of Douglas fir-larch or southern pine may be used to resist shear due to wind or seismic forces not exceeding 300 pounds per lineal foot of width. The allowable strength shall be adjusted by the factors 0.82 and 0.65 where nails are used with sheathing and framing of Group III or IV wood species as listed in Table No. 25-17-J of U.B.C. Standard No. 25-17.

2. **Special construction.** Special diagonally sheathed diaphragms shall conform to conventional construction and in addition shall have all elements designed in conformance with the provisions of this code.

Each chord or portion thereof may be considered as a beam loaded with a uniform load per foot equal to 50 percent of the unit shear due to diaphragm action. The load shall be assumed as acting normal to the chord, in the plane of the diaphragm and either toward or away from the diaphragm. The span of the chord, or portion thereof, shall be the distance between structural members of the diaphragm, such as the joists, studs and blocking, which serve to transfer the assumed load to the sheathing.

Special diagonally sheathed diaphragms shall include conventional diaphragms sheathed with two layers of diagonal sheathing at 90 degrees to each other and on the same face of the supporting members.

Special diagonally sheathed diaphragms of Douglas fir-larch or southern pine may be used to resist shears due to wind or seismic loads, provided such shears do not stress the nails beyond their allowable safe lateral strength and do not exceed 600 pounds per lineal foot of width. The allowable strength shall be adjusted by the factors 0.82 and 0.65 where nails are used with sheathing and framing of Group III or IV wood species as listed in Table No. 25-17-J of U.B.C. Standard No. 25-17.

(c) **Plywood Diaphragms.** Horizontal and vertical diaphragms sheathed with plywood may be used to resist horizontal forces not exceeding those set forth in Table No. 25-J for horizontal diaphragms and Table No. 25-K for vertical diaphragms, or may be calculated by principles of mechanics without limitation by using values of nail strength and plywood shear values as specified elsewhere in this code. Plywood for horizontal diaphragms shall be as set forth in Table No. 25-R for corresponding joist spacing and loads. Plywood in shear walls shall be at least $\frac{5}{16}$ inch thick for studs spaced 16 inches on center and $\frac{3}{8}$ inch thick where studs are spaced 24 inches on center.

Maximum spans for plywood subfloor underlayment shall be as set forth in Table No. 25-S. Plywood used for horizontal and vertical diaphragms shall conform to U.B.C. Standard No. 25-9.

All boundary members shall be proportioned and spliced where necessary to transmit direct stresses. Framing members shall be at least 2-inch nominal in the dimension to which the plywood is attached. In general, panel edges shall bear on the framing members and butt along their center lines. Nails shall be placed not less than $\frac{3}{8}$ inch in from the panel edge, shall be spaced not more than 6 inches on center along panel edge bearings, and shall be firmly driven into the framing members. No unblocked panels less than 12 inches wide shall be used.

Fiberboard Sheathing Diaphragms

Sec. 2515. Wood stud walls sheathed with fiberboard sheathing complying with U.B.C. Standard No. 25-24 may be used to resist horizontal forces not exceeding those set forth in Table No. 25-0. The fiberboard sheathing, 4 feet by 8 feet, shall be applied vertically to wood studs not less than 2-inch nominal in thickness spaced 16 inches on center. Nailing shown in Table No. 25-0 shall be provided at the perimeter of the sheathing board and at intermediate studs. Blocking not less than 2-inch nominal in thickness shall be provided at horizontal joints when wall height exceeds length of sheathing panel, and sheathing shall be fastened to the blocking with nails sized as shown in Table No. 25-O spaced 3 inches on centers each side of joint. Nails shall be spaced not less than $\frac{3}{8}$ inch from edges and ends of sheathing. Marginal studs of shear walls or shear-resisting elements shall be adequately anchored at top and bottom and designed to resist all forces. The maximum height-width ratio shall be one and one-half to one.

2518 (g)

5. **Bracing.** All exterior walls and main cross stud partitions shall be effectively and thoroughly braced at each end, or as near thereto as possible, and at least every 25 feet of length by one of the following methods:

   A. Nominal 1-inch by 4-inch continuous diagonal braces let into top and bottom plates and intervening studs, placed at an angle not more than 60 degrees nor less than 45 degrees from the horizontal, and attached to the framing in conformance with Table No. 25-P.

   B. Wood boards of $\frac{5}{8}$-inch net minimum thickness applied diagonally on studs spaced not over 24 inches on center.

   C. Plywood sheathing with a thickness not less than $\frac{5}{16}$ inch for 16-inch stud spacing and not less than $\frac{3}{8}$ inch for 24-inch stud spacing in accordance with Tables No. 25-M and No. 25-N.

   D. Fiberboard sheathing 4-foot by 8-foot panels not less than $\frac{7}{16}$ inch thick applied vertically on studs spaced not over 16 inches on center when installed in accordance with Section 2515 and Table No. 25-0.

   E. Gypsum sheathing panels not less than $\frac{1}{2}$ inch thick on studs spaced not over 16 inches on center when installed in accordance with Table No. 47-I.

F. Particleboard Exterior Type 2-B-1 sheathing panels not less than ⅜ inch thick on studs spaced not more than 16 inches on center.

G. Gypsum wallboard not less than ½ inch thick on studs spaced not over 24 inches on center when installed in accordance with Table No. 47-I.

H. Portland cement plaster on studs spaced 16 inches on center installed in accordance with Table No. 47-I.

For methods B, C, D, E, F, G and H, the braced panel must be at least 48 inches in width, covering three stud spaces where studs are spaced 16 inches apart and covering two stud spaces where studs are spaced 24 inches apart.

Solid sheathing of one of the materials specified in Items B through F, gypsum wallboard in Item G applied to supports at 16 inches on center, or portland cement plaster in Item H shall be applied to the exterior walls of the first story of all wood framed buildings three stories in height. In Seismic Zones Nos. 3 and 4 such braced wall sections shall be located at each end, or as near thereto as possible, and shall comprise at least 40 percent of the linear length of the wall.

Solid sheathing of one of the materials specified in Items B through F, gypsum wallboard in Item G applied to supports at 16 inches on center, or portland cement plaster in Item H shall be applied on either face of the exterior walls of the first story of all wood framed, two-story buildings and the second story of three-story buildings located in Seismic Zones No. 3 and No. 4. Braced wall sections shall be located at each end or as near thereto as possible and comprise at least 25 percent of the linear length of the wall.

All vertical joints of panel sheathing shall occur over studs. Horizontal joints shall occur over blocking equal in size to the studding except where waived by the installation requirements for the specific sheathing materials.

TABLE NO. 25-I—MAXIMUM DIAPHRAGM DIMENSION RATIOS

MATERIAL	HORIZONTAL DIAPHRAGMS Maximum Span-Width Ratios	VERTICAL DIAPHRAGMS Maximum Height-Width Ratios
1. Diagonal sheathing, conventional	3:1	2:1
2. Diagonal sheathing, special	4:1	3½:1
3. Plywood, nailed all edges	4:1	3½:1
4. Plywood, blocking omitted at intermediate joints	4:1	2:1

TABLE NO. 25-J—ALLOWABLE SHEAR IN POUNDS PER FOOT FOR HORIZONTAL PLYWOOD DIAPHRAGMS WITH FRAMING OF DOUGLAS FIR-LARCH OR SOUTHERN PINE[1]

PLYWOOD GRADE	Common Nail Size	Minimum Nominal Penetration in Framing (In Inches)	Minimum Nominal Plywood Thickness (In Inches)	Minimum Nominal Width of Framing Member (In Inches)	BLOCKED DIAPHRAGMS — Nail spacing at diaphragm boundaries (all cases), at continuous panel edges parallel to load (Cases 3 and 4) and at all panel edges (Cases 5 and 6).				UNBLOCKED DIAPHRAGM — Nails spaced 6" max. at supported end	
					6	4	2½	2	Load perpendicular to unblocked edges and continuous panel joints (Case 1)	Other configurations (Cases 2, 3 & 4)
					Nail spacing at other plywood panel edges					
					6	6	4	3		
STRUCTURAL I	6d	1¼	5/16	2	185	250	375	420	165	125
				3	210	280	420	475	185	140
	8d	1½	3/8	2	270	360	530	600	240	180
				3	300	400	600	675	265	200
	10d	1⅝	1/2	2	320	425	640[2]	730[2]	285	215
				3	360	480	720	820	320	240
C-D, C-C, STRUCTURAL II and other grades covered in U.B.C. Standard No. 25-9	6d	1¼	5/16	2	170	225	335	380	150	110
				3	190	250	380	430	170	125
			3/8	2	185	250	375	420	165	125
				3	210	280	420	475	185	140
	8d	1½	3/8	2	240	320	480	545	215	160
				3	270	360	540	610	240	180
			1/2	2	270	360	530	600	240	180
				3	300	400	600	675	265	200
	10d	1⅝	1/2	2	290	385	575[2]	655[2]	255	190
				3	325	430	650	735	290	215
			5/8	2	320	425	640[2]	730[2]	285	215
				3	360	480	720	820	320	240

¹These values are for short time loads due to wind or earthquake and must be reduced 25 percent for normal loading. Space nails 10 inches on center for floors and 12 inches on center for roofs along intermediate framing members.

Allowable shear values for nails in framing members of other species set forth in Table No. 25-17-J of U.B.C. Standards shall be calculated for all grades by multiplying the values for nails in STRUCTURAL I by the following factors: Group III, 0.82 and Group IV, 0.65.

²Reduce tabulated allowable shears 10 percent when boundary members provide less than 3-inch nominal nailing surface.

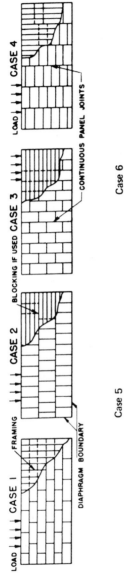

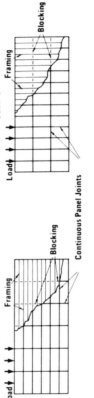

NOTE: Framing may be located in either direction for blocked diaphragms.

301

TABLE NO. 25-K—ALLOWABLE SHEAR FOR WIND OR SEISMIC FORCES IN POUNDS PER FOOT FOR PLYWOOD SHEAR WALLS WITH FRAMING OF DOUGLAS FIR-LARCH OR SOUTHERN PINE[1]

PLYWOOD GRADE	MINIMUM NOMINAL PLYWOOD THICKNESS (Inches)	MINIMUM NAIL PENETRATION IN FRAMING (Inches)	NAIL SIZE (Common or Galvanized Box)	PLYWOOD APPLIED DIRECT TO FRAMING — Nail Spacing at Plywood Panel Edges				NAIL SIZE (Common or Galvanized Box)	PLYWOOD APPLIED OVER ½-INCH GYPSUM SHEATHING — Nail Spacing at Plywood Panel Edges			
				6	4	2½	2		6	4	2½	2
STRUCTURAL I	5/16	1¼	6d	200	300[3]	450[3]	510	8d	200	300	450[2]	510
	3/8	1½	8d	230[3]	360[3]	530[3]	610[3]	10d	280	430	640[2]	730[2]
	1/2	1⅝	10d	340	510	770[2]	870[2]	—	—	—	—	—
C-D, C-C, STRUCTURAL II and other grades covered in U.B.C. Standard No. 25-9	5/16	1¼	6d	180	270	400	450	8d	180	270	400	450
	3/8	1½	8d	220[3]	320[3]	470[3]	530[3]	10d	260	380	570[2]	640[2]
	1/2	1⅝	10d	310	460	690[2]	770[2]	—	—	—	—	—
			NAIL SIZE (Galvanized Casing)					NAIL SIZE (Galvanized Casing)				
Plywood Panel Siding in Grades Covered in U.B.C. Standard No. 25-9	5/16	1¼	6d	140	210	320	360	8d	140	210	320	360
	3/8	1½	8d	130[3]	200[3]	300[3]	340[3]	10d	160	240	410	410

[1] All panel edges backed with 2-inch nominal or wider framing. Plywood installed either horizontally or vertically. Space nails at 6 inches on center along intermediate framing members for 3/8-inch plywood installed with face grain parallel to studs spaced 24 inches on center and 12 inches on center for other conditions and plywood thicknesses. These values are for short time loads due to wind or earthquake and must be reduced 25 percent for normal loading.

Allowable shear values for nails in framing members of other species set forth in Table No. 25-17-J of U.B.C. Standards shall be calculated for all grades by multiplying the values for common and galvanized box nails in STRUCTURAL I and galvanized casing nails in other grades by the following factors: Group III, 0.82 and Group IV, 0.65.

[2] Reduce tabulated allowable shears 10 percent when boundary members provide less than 3-inch nominal nailing surface.

[3] The values for 3/8-inch thick plywood applied direct to framing may be increased 20 percent, provided studs are spaced a maximum of 16 inches on center or plywood is applied with face grain across studs or if the plywood thickness is increased to ½ inch or greater.

TABLE NO. 25-O—ALLOWABLE SHEARS FOR WIND OR SEISMIC LOADING ON VERTICAL DIAPHRAGMS OF FIBERBOARD SHEATHING BOARD CONSTRUCTION FOR TYPE V CONSTRUCTION ONLY[1]

SIZE AND APPLICATION	NAIL SIZE	SHEAR VALUE 3-INCH NAIL SPACING AROUND PERIMETER AND 6-INCH AT INTERMEDIATE POINTS
$\frac{7}{16}$" x 4' x 8'	No. 11 ga. gal. roofing nail 1½" long, $\frac{7}{16}$" head	125[2]
$\frac{31}{32}$" x 4' x 8'	No. 11 ga. gal. roofing nail 1¾" long, $\frac{7}{16}$" head	175

[1]Fiberboard sheathing diaphragms shall not be used to brace concrete or masonry walls.

[2]The shear value may be 175 for ½-inch x 4 foot x 8 foot fiberboard nailbase sheathing.

Shear-resisting Construction with Wood Frame

Sec. 4713. (a) General. Portland cement plaster, gypsum lath and plaster, gypsum veneer base, gypsum sheathing board and gypsum wallboard may be used on wood studs for vertical diaphragms if applied in accordance with this section. Shear-resisting values shall not exceed those set forth in Table No. 47-I.

The shear values tabulated shall not be cumulative with the shear value of other materials applied to the same wall. The shear values may be doubled when the identical materials applied as specified in this section are applied to both sides of the wall.

(b) Masonry and Concrete Construction. Portland cement plaster, gypsum lath and plaster, gypsum veneer base, gypsum sheathing board and gypsum wallboard shall not be used in vertical diaphragms to resist forces imposed by masonry or concrete construction.

(c) Wall Framing. Framing for vertical diaphragms shall conform with Section 2518 (g) for bearing walls, and studs shall be spaced not further apart than 16 inches center to center. Marginal studs and plates shall be anchored to resist all design forces.

(d) Height-to-length Ratio. The maximum allowable height-to-length ratio for the construction in this section shall be 1½ to 1.

(e) Application. End joints of adjacent courses of gypsum lath, gypsum veneer base, gypsum sheathing board or gypsum wallboard sheets shall not occur over the same stud.

Where required in Table No. 47-I, blocking having the same cross-sectional dimensions as the studs shall be provided at all joints that are perpendicular to the studs.

The size and spacing of nails shall be as set forth in Table No. 47-I. Nails shall be spaced not less than ⅜ inch from edges and ends of gypsum lath, gypsum veneer base, gypsum sheathing board, gypsum wallboard or sides of studs, blocking and top and bottom plates.

1. **Gypsum lath.** Gypsum lath shall be applied perpendicular to the studs. Maximum allowable shear values shall be as set forth in Table No. 47-I.

2. **Gypsum sheathing board.** Four-foot-wide pieces may be applied parallel or perpendicular to studs. Two-foot-wide pieces shall be applied perpendicular to the studs. Maximum allowable shear values shall be as set forth in Table No. 47-I.

3. **Gypsum wallboard or veneer base.** Gypsum wallboard or veneer base may be applied parallel or perpendicular to studs. Maximum allowable shear values shall be as set forth in Table No. 47-I.

TABLE NO. 47-H—APPLICATION OF TWO-PLY GYPSUM WALLBOARD[1]

THICKNESS OF GYPSUM WALLBOARD (Each Ply) (inch)	PLANE OF FRAMING SURFACE	LONG DIMENSION OF GYPSUM WALLBOARD SHEETS	MAXIMUM SPACING OF FRAMING MEMBERS (Center to Center) (in Inches)	MAXIMUM SPACING OF FASTENERS (Center to Center) (in Inches)				
				Base Ply			Face Ply	
				Nails[2]	Screws[3]	Staples[4]	Nails[2]	Screws[3]
FASTENERS ONLY								
3/8	Horizontal	Perpendicular only	16	16	24	16	7	12
	Vertical	Either Direction	16				8	
1/2	Horizontal	Perpendicular only	24				7	
	Vertical	Either Direction	24				8	
5/8	Horizontal	Perpendicular only	24				7	
	Vertical	Either Direction	24				8	
Fasteners and Adhesives								
3/8 Base Ply	Horizontal	Perpendicular only	16	7	12	5	Temporary Nailing or Shoring to Comply with Section 4711 (d)	
	Vertical	Either Direction	24	8		7		
1/2 Base Ply	Horizontal	Perpendicular only	24	7		5		
	Vertical	Either Direction	24	8		7		
5/8 Base Ply	Horizontal	Perpendicular only	24	7		5		
	Vertical	Either Direction	24	8		7		

[1]For fire-resistive construction, see Tables No. 43-B and No. 43-C. For shear-resisting elements, see Table No. 47-I.

[2]Nails for wood framing shall be long enough to penetrate into wood members not less than 7/8 inch and the sizes shall conform with the provisions of Table No. 47-G. For nails not included in Table No. 47-G, use the appropriate size cooler nail as set forth in Table No. 25-17-1 of U.B.C. Standard No. 25-17. Nails for metal framing shall conform with the provisions of Table No. 47-G.

[3]Screws shall conform with the provisions of Table No. 47-G.

[4]Staples shall be not less than No. 16 gauge by 7/16-inch crown width with leg length of 7/8 inch, 1 1/8 inches and 1 3/8 inches for gypsum wallboard thicknesses of 3/8 inch, 1/2 inch and 5/8 inch, respectively.

TABLE NO. 47-I—ALLOWABLE SHEAR FOR WIND OR SEISMIC FORCES IN POUNDS PER FOOT FOR VERTICAL DIAPHRAGMS OF LATH AND PLASTER OR GYPSUM BOARD FRAME WALL ASSEMBLIES[1]

TYPE OF MATERIAL	THICKNESS OF MATERIAL	WALL CONSTRUCTION	NAIL SPACING MAXIMUM (In Inches)	SHEAR VALUE	MINIMUM NAIL SIZE
1. Expanded metal, or woven wire lath and portland cement plaster	⅞"	Unblocked	6	180	No. 11 gauge, 1½" long, ⁷⁄₁₆" head No. 16 gauge staple, ⅞" legs
2. Gypsum lath, plain or perforated	⅜" Lath and ½" Plaster	Unblocked	5	100	No. 13 gauge, 1⅛" long, ¹⁹⁄₆₄" head, plasterboard blued nail.
3. Gypsum sheathing board	½" x 2' x 8'	Unblocked	4	75	No. 11 gauge, 1¾" long, ⁷⁄₁₆" head, diamond-point, galvanized.
	½" x 4' ½" x 4'	Blocked Unblocked	4 7	175 100	
4. Gypsum wallboard or veneer base	½"	Unblocked	7	100	5d cooler nails.
			4	125	
		Blocked	7	125	
			4	150	
	⅝"	Blocked	4	175	6d cooler nails.
	⅝"	Blocked Two-ply	Base ply 9 Face ply 7	250	Base ply—6d cooler nails. Face ply—8d cooler nails.

[1]These vertical diaphragms shall not be used to resist loads imposed by masonry or concrete construction. See Section 4713 (b). Values are for short-time loading due to wind or earthquake and must be reduced 25 percent for normal loading.

[2]Applies to nailing at all studs, top and bottom plates and blocking.

Area Limitations

Sec. 5403. Exterior glass and glazing shall be capable of safely withstanding the loads set forth in Table No. 23-F, acting inward or outward. The area of individual lights shall be not more than set forth in Table No. 54-A or as adjusted by Table No. 54-B.

TABLE NO. 54-B—ADJUSTMENT FACTORS—RELATIVE RESISTANCE TO WIND LOAD[1]

GLASS TYPE	APPROXIMATE RELATIONSHIP
1. Laminated	0.6
2. Wired	0.5
3. Heat-strengthened	2.0
4. Fully tempered	4.0
5. Factory-fabricated Double Glazing[2]	1.5
6. Rough Rolled Plate	1.0
7. Sandblasted	Varies
8. Regular Plate, Float or Sheet	1.0

To determine the maximum allowable area for glass types listed in Table No. 54-B multiply the allowable area established in Table No. 54-A by the appropriate adjustment factor. Example: For ¼-inch heat-strengthened glass determine the maximum allowable area for a 30-pound-per-square-foot wind load requirement. Solution procedure: Use Table No. 54-A to determine the established allowable area for ¼-inch plate or float glass. Answer: 36 square feet, then multiply 36 by 2—the heat-strengthened glass adjustment factor. Answer: 72.

[2]Use thickness of the thinner of the two lights, not thickness of the unit.

[1]To be approved by the building official since adjustment factor varies with amount of depreciation and type of glass.

TABLE NO. 54-A—MAXIMUM ALLOWABLE AREA OF GLASS[1]
(In Square Feet)

WIND LOAD (In Pounds per Square Foot)	PLATE OR FLOAT GLASS THICKNESS (In Inches)													SHEET GLASS THICKNESS (In Inches)								
	1/8	3/16	13/64	7/32	1/4	5/16	3/8	1/2	5/8	3/4	7/8	1	1 1/4	SS	DS	3/16	7/32	1/4	11/32	3/8	7/16	1/2
10	41	72	81	89	107	144	185	275	351	465	525	656	956	41	56	95	109	128	186	213	243	311
15	27	48	54	60	71	96	123	183	234	310	350	438	637	27	38	63	73	86	124	142	162	207
20	21	36	40	45	53	72	92	137	176	232	262	328	478	20	28	47	55	64	93	107	122	155
25	16	29	32	36	43	58	74	110	140	186	210	262	382	16	23	38	44	51	74	85	97	124
30	14	24	27	30	36	48	62	92	117	155	175	219	319	14	19	32	36	43	62	71	81	104
35	12	21	23	26	31	41	53	79	100	133	150	188	273	12	16	27	31	37	53	61	69	89
40	10	18	20	22	27	36	46	69	88	116	131	164	239	10	14	24	27	32	46	53	61	78
45	9	16	18	20	24	32	41	61	78	103	117	146	212	9	13	21	24	29	41	47	54	69
50	8	14	16	18	21	29	37	55	70	93	105	131	191	8	11	19	22	26	37	43	49	62
60	7	12	13	15	18	24	31	46	59	77	88	109	159	7	9	16	18	21	31	36	41	52
70	6	10	12	13	15	21	26	39	50	66	75	94	137	6	8	14	16	18	27	30	35	44
80	5	9	10	11	13	18	23	34	44	58	66	82	120	5	7	12	14	16	23	27	30	39
90	4.5	8	9	10	12	16	21	31	39	52	58	73	106	4.5	6	11	12	14	21	24	27	35
100	4	7	8	9	11	14	18	27	35	46	52	66	96	4	5.5	9	11	13	19	21	24	31

[1]Maximum areas apply for rectangular lights of plate, float or sheet glass firmly supported on all four sides in a vertical position. Glass mounted at a slope not to exceed one horizontal to five verticals may be considered vertical. Maximum areas based on minimum thicknesses set forth in Table No. 54-1-C, Uniform Building Code Standard No. 54-1.

h/d	Rc	h/d	Rc	h/d	Rc	h/d	Rc	h/d	Rc	h/d	Rc
9.90	.0006	5.20	.0043	1.85	.0810	1.38	.1706	0.91	.4352	0.45	1.4582
9.80	.0007	5.10	.0046	1.84	.0821	1.37	.1737	0.90	.4452	0.44	1.5054
9.70	.0007	5.00	.0049	1.83	.0833	1.36	.1768	0.89	.4554	0.43	1.5547
9.60	.0007	4.90	.0052	1.82	.0845	1.35	.1800	0.88	.4659	0.42	1.6063
9.50	.0007	4.80	.0055	1.81	.0858	1.34	.1832	0.87	.4767	0.41	1.6604
9.40	.0007	4.70	.0058	1.80	.0870	1.33	.1866	0.86	.4899	0.40	1.7170
9.30	.0008	4.60	.0062	1.79	.0883	1.32	.1900	0.85	.4994	0.39	1.7765
9.20	.0008	4.50	.0066	1.78	.0896	1.31	.1935	0.84	.5112	0.38	1.8380
9.10	.0008	4.40	.0071	1.77	.0909	1.30	.1970	0.83	.5233	0.37	1.9098
9.00	.0008	4.30	.0076	1.76	.0923	1.29	.2007	0.82	.5359	0.36	1.9738
8.90	.0009	4.20	.0081	1.75	.0937	1.28	.2044	0.81	.5488	0.35	2.0467
8.80	.0009	4.10	.0087	1.74	.0951	1.27	.2083	0.80	.5621	0.34	2.1237
8.70	.0009	4.00	.0093	1.73	.0965	1.26	.2122	0.79	.5758	0.33	2.2051
8.60	.0010	3.90	.0100	1.72	.0980	1.25	.2162	0.78	.5899	0.32	2.2913
8.50	.0010	3.80	.0108	1.71	.0995	1.24	.2203	0.77	.6044	0.31	2.3828
8.40	.0010	3.70	.0117	1.70	.1010	1.23	.2245	0.76	.6194	0.30	2.4802
8.30	.0011	3.60	.0127	1.69	.1026	1.22	.2289	0.75	.6349	0.29	2.5838
8.20	.0012	3.50	.0137	1.68	.1041	1.21	.2333	0.74	.6509	0.28	2.6945
8.10	.0012	3.40	.0149	1.67	.1058	1.20	.2378	0.73	.6674	0.27	2.8130
8.00	.0012	3.30	.0163	1.66	.1074	1.19	.2425	0.72	.6844	0.26	2.9401
7.90	.0013	3.20	.0178	1.65	.1091	1.18	.2472	0.71	.7019	0.25	3.0769
7.80	.0013	3.10	.0195	1.64	.1108	1.17	.2521	0.70	.7200	0.24	3.2246
7.70	.0014	3.00	.0214	1.63	.1125	1.16	.2571	0.69	.7388	0.23	3.3845
7.60	.0014	2.90	.0235	1.62	.1143	1.15	.2622	0.68	.7581	0.22	3.5583
7.50	.0015	2.80	.0260	1.61	.1162	1.14	.2675	0.67	.7781	0.21	3.7479
7.40	.0015	2.70	.0288	1.60	.1180	1.13	.2729	0.66	.7987	0.20	3.9557
7.30	.0016	2.60	.0320	1.59	.1199	1.12	.2784	0.65	.8201	.195	4.0673
7.20	.0017	2.50	.0357	1.58	.1218	1.11	.2841	0.64	.8422	.190	4.1845
7.10	.0017	2.40	.0400	1.57	.1238	1.10	.2899	0.63	.8650	.185	4.3079
7.00	.0018	2.30	.0450	1.56	.1258	1.09	.2959	0.62	.8886	.180	4.4379
6.90	.0019	2.20	.0508	1.55	.1279	1.08	.3020	0.61	.9131	.175	4.5751
6.80	.0020	2.10	.0577	1.54	.1300	1.07	.3083	0.60	.9384	.170	4.7201
6.70	.0020	2.00	.0658	1.53	.1322	1.06	.3147	0.59	.9647	.165	4.8736
6.60	.0021	1.99	.0667	1.52	.1344	1.05	.3213	0.58	.9919	.160	5.0364
6.50	.0022	1.98	.0676	1.51	.1366	1.04	.3281	0.57	1.0201	.155	5.2095
6.40	.0023	1.97	.0685	1.50	.1389	1.03	.3351	0.56	1.0493	.150	5.3937
6.30	.0025	1.96	.0694	1.49	.1412	1.02	.3422	0.55	1.0797	.145	5.5904
6.20	.0026	1.95	.0704	1.48	.1436	1.01	.3496	0.54	1.1112	.140	5.8008
6.10	.0027	1.94	.0714	1.47	.1461	1.00	.3571	0.53	1.1439	.135	6.0261
6.00	.0028	1.93	.0724	1.46	.1486	0.99	.3649	0.52	1.1779	.130	6.2696
5.90	.0030	1.92	.0734	1.45	.1511	0.98	.3729	0.51	1.2132	.125	6.5306
5.80	.0031	1.91	.0744	1.44	.1537	0.97	.3811	0.50	1.2500	.120	6.8136
5.70	.0033	1.90	.0754	1.43	.1564	0.96	.3895	0.49	1.2883	.115	7.1208
5.60	.0035	1.89	.0765	1.42	.1591	0.95	.3981	0.48	1.3281	.110	7.4555
5.50	.0037	1.88	.0776	1.41	.1619	0.94	.4070	0.47	1.3696	.105	7.8215
5.40	.0039	1.87	.0787	1.40	.1647	0.93	.4162	0.46	1.4130	.100	8.2237
5.30	.0041	1.86	.0798	1.39	.1676	0.92	.4255				

DISTRIBUTION OF HORIZONTAL FORCES ALONG A FIXED MASONRY WALL—AVERAGE RIGIDITIES

h/d	R_f	h/d	R_f	h/d	R_f	h/d	R_f	h/d	R_f	h/d	R_f
9.90	.0025	5.20	.0160	1.85	.2104	1.38	.3694	0.91	.7177	0.45	1.736
9.80	.0026	5.10	.0169	1.84	.2128	1.37	.3742	0.90	.7291	0.44	1.779
9.70	.0027	5.00	.0179	1.83	.2152	1.36	.3790	0.89	.7407	0.43	1.825
9.60	.0027	4.90	.0189	1.82	.2176	1.35	.3840	0.88	.7527	0.42	1.874
9.50	.0028	4.80	.0200	1.81	.2201	1.34	.3890	0.87	.7649	0.41	1.924
9.40	.0029	4.70	.0212	1.80	.2226	1.33	.3942	0.86	.7773	0.40	1.978
9.30	.0030	4.60	.0225	1.79	.2251	1.32	.3994	0.85	.7901	0.39	2.034
9.20	.0031	4.50	.0239	1.78	.2277	1.31	.4047	0.84	.8031	0.38	2.092
9.10	.0032	4.40	.0254	1.77	.2303	1.30	.4100	0.83	.8165	0.37	2.154
9.00	.0033	4.30	.0271	1.76	.2330	1.29	.4155	0.82	.8302	0.36	2.219
8.90	.0034	4.20	.0288	1.75	.2356	1.28	.4211	0.81	.8442	0.35	2.287
8.80	.0035	4.10	.0308	1.74	.2384	1.27	.4267	0.80	.8585	0.34	2.360
8.70	.0037	4.00	.0329	1.73	.2411	1.26	.4324	0.79	0.873	0.33	2.437
8.60	.0038	3.90	.0352	1.72	.2439	1.25	.4384	0.78	0.888	0.32	2.518
8.50	.0039	3.80	.0377	1.71	.2468	1.24	.4443	0.77	0.904	0.31	2.605
8.40	.0040	3.70	.0405	1.70	.2497	1.23	.4504	0.76	0.920	0.30	2.697
8.30	.0042	3.60	.0435	1.69	.2526	1.22	.4566	0.75	0.936	0.29	2.795
8.20	.0043	3.50	.0468	1.68	.2556	1.21	.4628	0.74	0.952	0.28	2.900
8.10	.0045	3.40	.0505	1.67	.2586	1.20	.4692	0.73	0.969	0.27	3.013
8.00	.0047	3.30	.0545	1.66	.2617	1.19	.4757	0.72	0.987	0.26	3.135
7.90	.0048	3.20	.0590	1.65	.2648	1.18	.4823	0.71	1.005	0.25	3.265
7.80	.0050	3.10	.0640	1.64	.2679	1.17	.4891	0.70	1.023	0.24	3.407
7.70	.0052	3.00	.0694	1.63	.2711	1.16	.4959	0.69	1.042	0.23	3.560
7.60	.0054	2.90	.0756	1.62	.2744	1.15	.5029	0.68	1.062	0.22	3.728
7.50	.0056	2.80	.0824	1.61	.2777	1.14	.5100	0.67	1.082	0.21	3.911
7.40	.0058	2.70	.0900	1.60	.2811	1.13	.5173	0.66	1.103	0.20	4.112
7.30	.0061	2.60	.0985	1.59	.2844	1.12	.5247	0.65	1.124	.195	4.220
7.20	.0063	2.50	.1081	1.58	.2879	1.11	.5322	0.64	1.146	.190	4.334
7.10	.0065	2.40	.1189	1.57	.2914	1.10	.5398	0.63	1.168	.185	4.454
7.00	.0069	2.30	.1311	1.56	.2949	1.09	.5476	0.62	1.191	.180	4.580
6.90	.0072	2.20	.1449	1.55	.2985	1.08	.5556	0.61	1.216	.175	4.714
6.80	.0075	2.10	.1607	1.54	.3022	1.07	.5637	0.60	1.240	.170	4.855
6.70	.0078	2.00	.1786	1.53	.3059	1.06	.5719	0.59	1.266	.165	5.005
6.60	.0081	1.99	.1805	1.52	.3097	1.05	.5804	0.58	1.292	.160	5.164
6.50	.0085	1.98	.1824	1.51	.3136	1.04	.5889	0.57	1.319	.155	5.334
6.40	.0089	1.97	.1844	1.50	.3175	1.03	.5977	0.56	1.347	.150	5.514
6.30	.0093	1.96	.1864	1.49	.3214	1.02	.6066	0.55	1.376	.145	5.707
6.20	.0098	1.95	.1885	1.48	.3245	1.01	.6157	0.54	1.407	.140	5.914
6.10	.0102	1.94	.1905	1.47	.3295	1.00	.6250	0.53	1.438	.135	6.136
6.00	.0107	1.93	.1926	1.46	.3337	0.99	.6344	0.52	1.470	.130	6.374
5.90	.0112	1.92	.1947	1.45	.3379	0.98	.6441	0.51	1.504	.125	6.632
5.80	.0118	1.91	.1969	1.44	.3422	0.97	.6540	0.50	1.539	.120	6.911
5.70	.0124	1.90	.1991	1.43	.3465	0.96	.6641	0.49	1.575	.115	7.215
5.60	.0130	1.89	.2013	1.42	.3510	0.95	.6743	0.48	1.612	.110	7.545
5.50	.0137	1.88	.2035	1.41	.3555	0.94	.6848	0.47	1.651	.105	7.908
5.40	.0144	1.87	.2058	1.40	.3600	0.93	.6955	0.46	1.692	.100	8.306
5.30	.0152	1.86	.2081	1.39	.3647	0.92	.7065				

310

1½" B (24")
ROOF DECK OR ACOUSTIDECK®
Painted and Galvanized

3 Puddle Welds

Side Seams
Button-Punched 36" O.C.

GAGE		SPANS					
		5' - 0"	6' - 0"	7' - 0"	8' - 0"	9' - 0"	10' - 0"
22	q	280	240	210	185	170	155
	F	23.84 + 224.5R	27.87 + 187.0R	31.83 + 160.3R	35.73 + 140.3R	39.56 + 124.7R	43.32 + 112.2R
20	q	445	375	330	295	270	245
	F	16.88 + 129.7R	19.88 + 108.1R	22.91 + 92.6R	25.94 + 81.0R	28.97 + 72.0R	32.01 + 64.8R
18	q	790	665	580	515	465	425
	F	10.44 + 54.9R	12.38 + 45.8R	14.38 + 39.2R	16.43 + 34.3R	18.52 + 30.5R	20.66 + 27.5R
16	q	1115	940	815	725	650	590
	F	7.23 + 28.1R	8.60 + 23.4R	10.03 + 20.0R	11.52 + 17.5R	13.06 + 15.6R	14.65 + 14.0R

q = diaphragm shear load in lb/ft

Seismic and Hurricane Anchors

for trusses and rafters

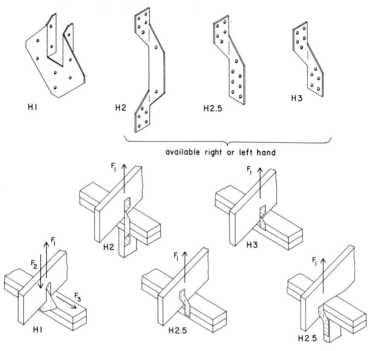

available right or left hand

No.	Fasteners			Load Capacity - lb		
	to rafters	to plates	to studs	F_1	F_2	F_3
H1	4 - 8d	4 - 8d	—	520	390	520
H2	5 - 8d	—	5 - 8d	370	—	—
H2.5	5 - 8d	4 to 6-8d	4 to 6-8d	370	—	—
H3	4 - 8d	4 - 8d	—	305	—	—

Holdowns

HD2

HD5

HD7

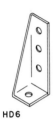

HD6

HD2N
HD5N
HD7N

Typical Installation Details

7D

length
as required
for load

7D

No.	Bolts		Design Load – in lbs (Normal Loading) with stud sizes shown		
	base	stud	1 9/16	2 9/16	3 9/16
HD2 HD2N	5/8	2 – 5/8	2450	2520	2520
HD5 HD5N	3/4	2 – 3/4	3375	3610	3610
HD6	1	3 – 3/4	5060	5410	5410
HD7	1	3 – 7/8	6350	6500	6500
	1 1/8	3 – 7/8	6350	7380	7380
HD7N	1	2 – 1	6350	6500	6500

313

Index